GW00702406

A Gourmet Odyssey
Noosa to Mooloolaba

A Gourmet Odyssey

NOOSA TO MOOLOOLABA

Petra Frieser

Little Hills Press

...dedicated to anyone prepared
to follow their dreams!

A Gourmet Odyssey
Noosa to Mooloolaba
By Petra Frieser

ISBN 1863152520

Designed and conceived by
Local Harvest.

Little Hills Press
Unit 12, 103 Kurrajong Avenue
Mount Druitt NSW 2770
Australia
www.littlehills.com

Published by Little Hills Press 2005

Printed in China through Colorcraft Ltd, Hong Kong

www.localharvest.com.au

Noosa Granite Bay

A Gourmet Odyssey
NOOSA TO MOOLOOLABA
Table of Contents

A Gourmet Odyssey

NOOSA TO MOOLOOLABA

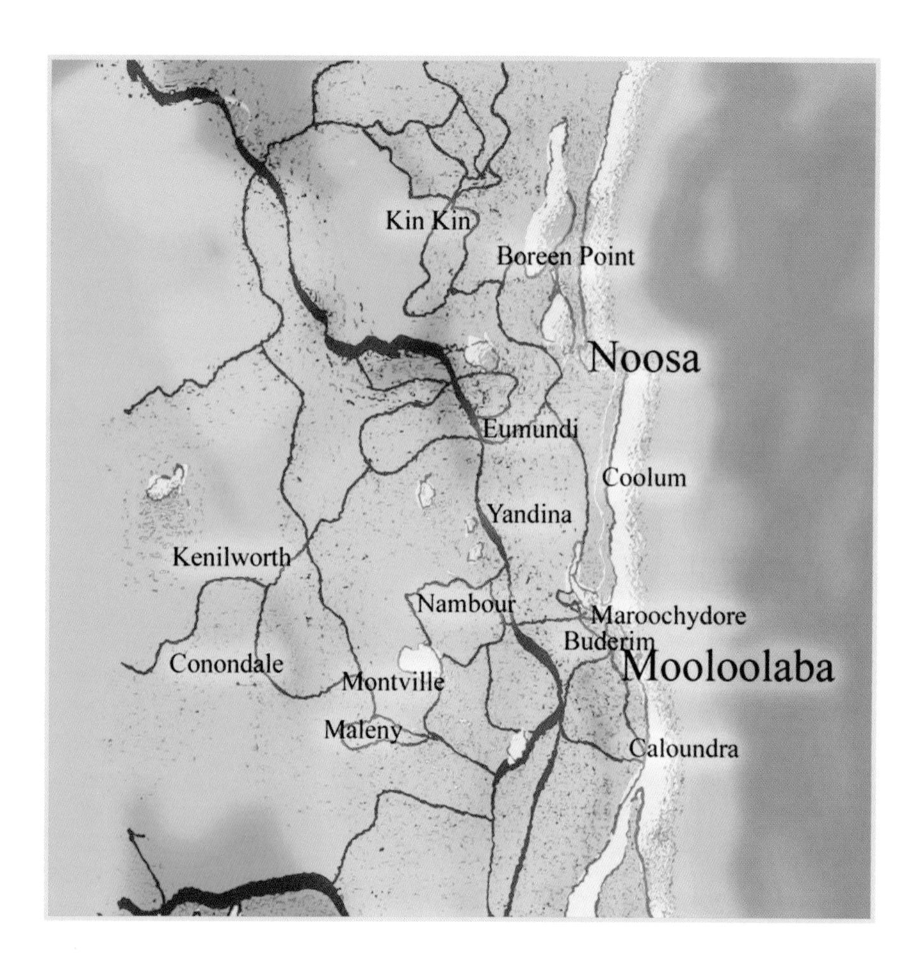

The Sunshine Coast needs no introduction when it comes to its pristine beaches and sapphire blue seas, what probably does need introducing is the cultural fabric that weaves a constant thread along this spectacular strip of eastern coastline with Noosa and Mooloolaba sparkling brightly as the brilliant jewels in the coastal crown.

The uniting thread in this instance is life sustaining, or perhaps more appropriately, lifestyle sustaining, as it would seem on the Sunshine Coast it is as much about lifestyle as it is anything else. There is a wealth of hidden treasure scattered throughout the Sunshine Coast in the way of innovative regional produce, many producers pioneers in their field, often taking enormous leaps of faith by planting crops that large sectors of the community have never even heard of, with results that are nothing short of spectacular.

There are so many jewel like facets to this industry, where at first you could be forgiven for thinking that all the Sunshine Coast can boast is endless crops of sugar cane and pineapples, nothing could in fact be further from the truth – diversity being the key. The climate lends itself to a large range of experimental sub-tropical and exotic crops and we continue to raise our interest in our own native foods, which will no doubt see a lot of growth commercially over the next decade. Wineries, a relatively young industry are harvesting their first crops, which will see the production of top quality wines in the very near future. Aquaculture industries thrive, providing local restaurants with numerous native fish species and red claw crayfish as well as having Mooloolaba as a main port of call for bounty laden fishing vessels that unload volumes of fresh tuna and other sought after seafood, that is then exported all over the world. There is much in the way of livestock – goats and cattle for both milk and meat, deer, poultry and emus, which has sparked curiosity of some of the more adventurous chefs who are keen to explore the merits of this flavoursome new meat. Then there are the multitude of other dedicated individuals who ply these base products into fabulous preserves in an attempt to add value to these already burgeoning industries.

It is with this in mind that a gourmet odyssey begins. Discover and explore the spectacular Sunshine Coast and unearth its valuable edible treasures. You don't need to be a connoisseur to appreciate the richness of flavour that each product has to offer, or be a chef to bring them to the table. A morning can be spent scouring the infamous Eumundi Markets where many smaller producers sell and allow you to sample their goods – or a drive through the ranges stopping at the various wineries and local farms can be equally fruitful, not to mention enjoyable.

And so, it would give me great pleasure to discover that, through your own exploration and experimentation, the gigantic lid was lifted, to expose the wonderful hidden treasure, that the Sunshine Coast has and is and discover and support the fabulous food and beverage industries of this Coast.

Noosa

If there were indeed a 'foodies' heaven, my guess it would be located somewhere in Noosa. Noosa has been the forerunner for fine dining for decades, chefs introducing an innovative and eclectic mix of regional fare to forge a Mediterranean, come Asian cuisine style, distinctly its own. Cosmopolitan cafés, brasseries and restaurants line Hastings Street and the fabulous Noosa River and various watercraft ferry anticipating diners to their waterfront destinations.

Noosa has it all. It is home to wonderful food festivals, a mesmerizing Farmer's Market along with a multitude of other food, music and art related events scattered throughout the year.

Noosa has a simplistic decadence, something of a contradiction in itself. Food is simple, raw and natural while being complicated, rich and luxurious all at once - everything in Noosa seems to reflect this same unique quality.

A visitor will always begin their journey on Hastings Street, beachfront establishments the definite favourite. The Noosa River follows close behind with food industry icons recognising the beauty of the waterfront, as the ideal position for the consumption of the finest of whatever Noosa has to offer. Noosa restaurants have always been great supporters of their local industries and menus have never been afraid to share this information with their diners. Many local producers have created their own niche markets supplying restaurants direct, with gloriously fresh Asian greens, herbs, sun ripened tomatoes and a range of organically grown fresh fruit and vegetables. A number of restaurants go that step further and incorporate seasonal organic and free-range produce wherever possible. Local free range poultry, game, beef and lamb, all find their way onto menus in the form of unforgettable gourmet experiences. Along side there sits a gorgeous abundance of fresh seafood, as is expected of a seaside community.

SURF RESCUE

HEY BILL!
HEY BILL'S
BICYCLE
HIRE
LP-436
ITALIANO
PILU
RISTORANTE BAR
la lista dei vini
SIERRA
CAFE BAR
SIERRA
CAFE BAR
SIERRA
CAFE BAR
berardo's
welcome
degustation
today's menu
upcoming events
fusions

16
HASTINGS STREET
parc food
HAPPY HOURS
5PM-6PM
9PM-10PM
TAP BEER from $2
BASIC SPIRITS 4
WINE 4
NACHOS 10
OYSTERS 15
TAPAS PLATTER 18
COCKTAILS 8
449 7441
RIVER HOU E REST
LINDONI'S
RISTORANTE
ITALIANO
RICKY RICARDO'S

There is large focus on the exotic in Noosa, especially of the cornucopia of exotic fruits. If there was such a thing as a tropical Garden of Eden, I am sure that the Noosa hinterland would be growing most of the garden's most prized possessions. Carambolas, black sapotes, custard apples and sapodillas – they are not fruits that many of us are familiar with, yet worthy of paradise none-the-less. These fascinating aberrations of nature, with their prehistorically textured skins and nectareous pulps are so exotically varied and obscurely structured that it cannot be said that any one is in anyway like to another. A favourite has long been the carambola; it is not quite so prehistoric in appearance and otherwise known as the starfruit. Its celestial attributes outshine its somewhat more subtle taste but it remains a refreshing novelty in salads and fruit desserts. The black sapote or chocolate pudding fruit is something of a wonder. It has a consistency as the name applies, of chocolate pudding. The taste isn't quite as convincing but the fruit is intriguing enough to warrant experimentations with flavourings such as rum or lemon juice for wonderful results. That aside, it is about four times richer in Vitamin C than an orange so certainly a healthier alternative to chocolate pudding! Custard apples, again as the name implies, tastes like a cross between custard and apples (or pineapples and bananas!). It is a great addition to smoothies and ice creams. Sapodillas - caramel springs to mind; abius – butterscotch; rollinias - lemon meringue... give a diverse and unique spectrum of flavours that remained relatively undiscovered by consumers.

The wonderful thing about the Sunshine Coast is that many people have at least had a try at growing some of the more exotic varieties of fruit. Some more seriously than others which has resulted in lonely laneways and lesser known hinterland roads being lined with the most amazing road side stalls, complete with honesty boxes and varieties of fruit that are simply not available in stores. A drive in the Noosa hinterland can indeed be a fruitful journey. If all else fails then a trip to either the **Noosa Farmers Market** or the **Eumundi Markets** is definitely in order.

Exotic fruit is something that needs to be tried. Some of the fruits are an acquired taste, but some are such paradise to the taste buds it won't take long to acquire a taste for them at all.

Chilli Jam Prawns

Chilli Jam Prawns are great for the BBQ. A variety of Chilli Jams are made by numerous local producers, tracking one down at the local markets shouldn't be hard – if not just substitute with a sweet chilli sauce.

1kg green prawns

¼ cup Chilli Jam

1 clove garlic - crushed

1 bunch coriander – roughly chopped

¼ tsp. freshly grated ginger

1 tbs. olive oil

Soak some bamboo skewers in hot water and set aside until ready to use. Remove the shells, heads and veins from the prawns and set aside.

Mix together the remaining ingredients and ¾ of the coriander in a bowl and then pour over the prawns. Marinate the prawns in the fridge for 1 hour. Place 3-4 prawns on each skewer.

When ready to serve place on a hot BBQ for a few minutes on each side. Serve hot with lemon or lime wedges and sprinkled with remaining fresh coriander.

Chillies

There are dozens of chilli varieties ranging from sweet to blazing hot. Chillies are the base ingredient to many curry pastes and Asian inspired recipes.

Look out for:

- Fresh Chillies
- Chilli Jam
- Chilli Beer
- Curry Pastes
- Strawberry Chilli Sauce
- Sweet Chilli Dipping Sauce
- Chilli Ice Cream

Noosa Hinterland

The place to begin a gourmet experience on the Sunshine Coast would have to be the **Eumundi Markets**. The Eumundi Markets is the star of the Sunshine Coast's gourmet universe, attracting thousands of visitors every weekend. It has by far the largest selection of fresh local produce and is the platform for experimentation and the introduction of innovative fare to the market place.

The market showcases a wide spectrum of goods and producers, growers of the various gourmet delights speckled throughout. Look no further for the freshest vegetables, many organically grown. Fruit not otherwise seen or available through normal vendors are now overflowing in a brilliant constellation of baskets and bundles. Preserves of every conceivable delicious combination of fruit or vegetable are available to try or buy. Freshly baked wood-fired bread, artisan cheeses, delightful beverages - it is a gourmand's dream come true. Wednesday markets have also become more dominant with the Old Heritage Market site now hosting a Farmers Market with it's own special selection of food producers side by side.

There are pestos made of astounding combinations of nuts and greens, soft cheeses and curds with herbs or olives, pickles and relishes that make the mouth-water. It would be unfair to single out individual producers as their products all shine equally, so the best advice is to grab a basket and a wad of money and spend either a Wednesday or Saturday morning shopping to your foodie heart's content!

Eumundi itself is a town with an enormous amount of tradition and charm. It is relatively undeveloped in the modern sense and locals are community minded in the preservation of the town's history and ambience. Eumundi's most recent residential addition is **The Australian Nougat Company**, Bliss….. they certainly knew the meaning of the word when the name for this little nougat making enterprise was developed. The 'Bliss' range is now huge – macadamia, pistachio, ginger, coffee and chocolate are some of the delectable flavours now on offer, all being made on the premises with, where ever possible, local produce.

EUMUNDI
THE OLD BAKERY
Est. 1909

Cow
MEZE GOURMET DELIGHT
Black Tapenade
MEZE GOURMET DELIGHT
Green Tapenade
Extra Virgin
Vintage 2002
Vintage 2002
SUE's
ORGANIC CUISINE
SUPPORTING G.E. FREE PRODUCTS
Handmade in Australia
Chilli Tomato Relish (Hot)
SUE's
ORGANIC CUISINE
SUPPORTING G.E. FREE PRODUCTS
& ONION JAM
275g Net
GYMPIE farm

now available!
Extra Virgin
Olive Oil
Extra Virgin
Olive Oil
ing oil or dressing
il & Garlic
Wild Grass
STUFFING
Sour Dough, Fig and Pistachio
Wild Grass
CALLY
INE
JAMS
DIPS

Ruby Blu
Grapefru
2 for $1

EUMUNDI MARKET LUNCH

Eumundi Market Lunch

This is what I call my 'Post Eumundi Market Lunch'. You can create any number of combinations, depending on what you have brought home. The Eumundi Markets are rife with pestos, tapenades and cheeses so it shouldn't be hard to source a mouth-watering combination.

For each person:

1 slice of Olive and Fetta Bread – cut nice and thick

1 large Mushroom

2 tbs. Hommus

1 tbs. Tapenade

¼ cup Kaffir Lime Oil

¼ cup Olive Oil

1 slice Goat's Cheese

1 small handful baby greens – bok choy or spinach

1 wedge of lime

cracked pepper

sprig of parsley

Put the olive oil and kaffir lime oil in a dish and marinate the whole mushroom in it while you are preparing the other ingredients.

Grill the mushrooms on the BBQ or place under the grill in the oven for 5-10 minutes. Toast a thick slice of bread under the grill/on the BBQ as well. Place the toast on a plate and spread with hommus. Place grilled mushroom on top. Plonk on the tapenade, baby greens and drizzle with the juices from the grill or the left over marinade.

Sprinkle with crumbled pieces of cheese and serve with a wedge of lime. Enjoy!

Between the strawberry farm just out of town, the **Eumundi Winery** on the other side of town and the treats from within the town itself, even a non-market day should prove fairly fruitful. The scenery is magic, ginger fields, strawberry fields and then the old Queenslander styled houses perched on vantage points in between.

Following Food Trails are a fabulous way of discovering a region. Local councils and tourism groups have worked together to establish these to offer farm visits for people wishing to gain a more personal insight to artisan food industries. The opportunity is there to see how things are grown or made and to buy seasonal produce directly from the farmer, or in some instances, pick your own produce fresh from the fields. The trails also allow visitors to explore lesser know regions of the Sunshine Coast with a culinary motive and destination in mind. All the Food Trails are picturesque, so from a visual perspective it doesn't really matter where you start.

One spectacular circuit, complete with winding roads and rolling hills is through the fertile hinterland region, which holds Boreen Point, Kin Kin and Wolvi in its tight embrace. Producers such as **Garnisha**, **Moran Group Teas, Yeltukka Pineapples** and **Jondell Macadamias** are based in the area, each contributing to the food industry with their own innovative range of produce.

Asian Greens & Tubers

Asian cuisine has inspired the growth of a multitude of zesty greens and tubers such as lemon grass, turmeric, galangal and kaffir limes. Markets are the best source for fresh produce.

Look out for:

- Earthcare Enterprises – seed and tubers for growing
- Curry Pastes
- Chutneys
- Lemon Grass Tea
- Kaffir Lime & Lemon Grass Oil

NOOSA
MASALA
250g NET
Made In Australia
GREEN
CURRY
Thai Style Curry Paste
GARNISHA
PRODUCE OF PARADISE
250g NET
Made In Australia

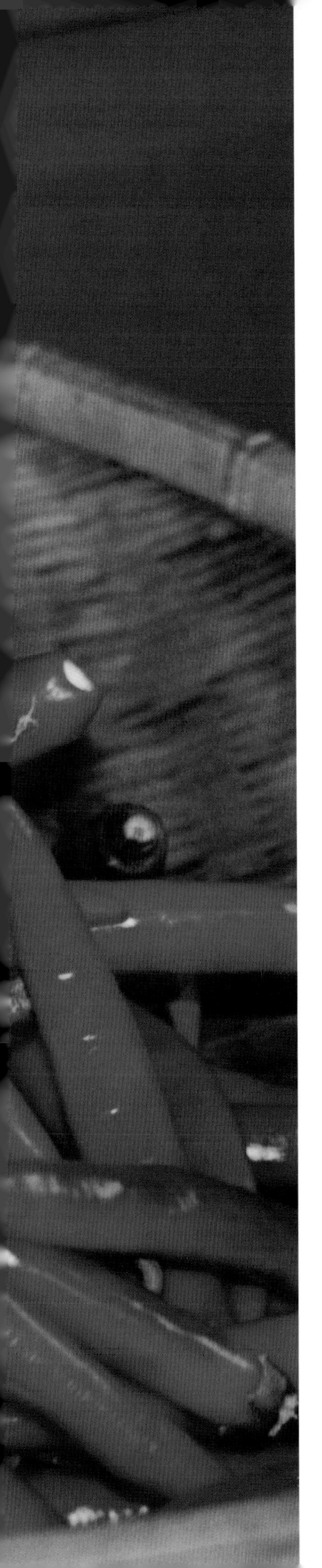

Thai Red Seafood Curry

THAI RED SEAFOOD CURRY

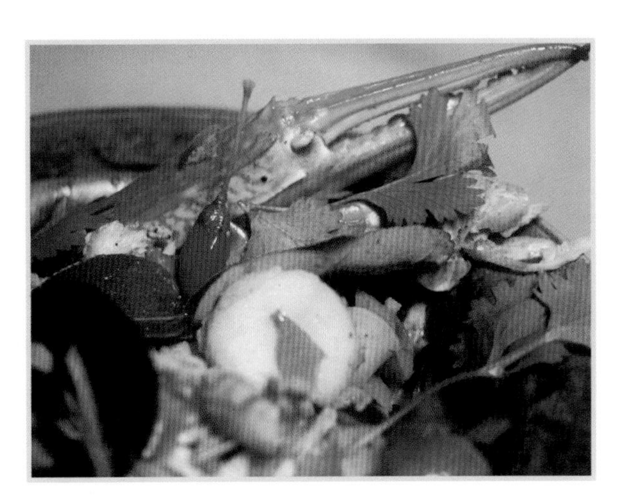

250g mussels

250g scallops

2 blue swimmer crabs

2 cuttlefish

3 tbs. Garnisha Red Curry Paste

375ml coconut milk

1 tbs. oil

1 clove garlic – crushed

2 shallots – chopped

1 lemon grass stalk – cut in half and bruised

4 kaffir lime leaves

2 birds eye chillies

fresh coriander sprigs

Rinse mussels and remove beards. Score the cuttlefish diagonally and cut into approximate 4cm squares. Break legs off the swimmer crab and set aside. Prise open the crab shell and gently scoop out the flesh and set aside also.

Heat the oil in a wok and add the shallots, garlic and curry paste and sauté for 1 minute. Add the coconut milk and simmer gently for 5 minutes. Add lemongrass, kaffir lime leaves and chillies. Add mussels, scallops, crabs legs and cuttlefish and simmer a further few minutes. Add crabmeat just before serving to heat through. When the mussels have all opened the curry should be ready to serve.

To serve, divide seafood between bowls and ladle curry juices over the top. Garnish with coriander and some chopped shallots. Serve with rice – stir a fingerlime through it, for a bit of extra zing.

Serves 4.

overleaf Jondell Macadamia Plantation

Macadamia Nuts

Macadamia nuts are incredibly versatile. Eaten raw or roasted, they are also processed into pastes and a beautifully nutty cholesterol free oil. The Sunshine Coast main processors are Suncoast Gold Macadamias, Jondell Macadamias and Nutworks.

Look out for:

- Macadamia Paste
- Macadamia Oil
- Macadamia Ice Cream
- Choc-coated Macadamia Bars
- Macadamia Nougat

Closest to Noosa is Boreen Point, a little township on the edge of Lake Cootharaba, a favourite camping and fishing destination and home to the lush tropical gardens of **Garnisha**. Garnisha is a short drive from town and is a hidden Utopia, brimming with exotic spices and fragrant greenery and tubers, that when ground, release the aromatic flavours which give the Garnisha curry pastes their piquant home made flavour. Growing on the property are alternate rows of lemongrass, galangal and cardamom alongside green mangos, curry plants and pimento trees amongst various other specimen style trees. A small olive grove and grapevines are full of promise of future harvests, as is a bountiful vegetable garden.

Garnisha specialises in Thai and Malay curry pastes, all of them with beautiful strong zesty flavours, available for purchase at the property or from select outlets throughout the coast. Farm Tours are offered where native meets tropical in a sensational, eclectic combination of regional produce.

On the outskirts of Kin Kin is **Moran Group Teas** or **Kin Kin Tea** as it is also known, producing an invigorating range of herbal teas with raw ingredients that are predominantly grown in the region's untainted soils. The Lemongrass tea is a favourite.The lemongrass is actually dried in a hut built on the property, which was modelled on a Malaysian Rice Drier. Drying usually takes about 1½ days and in the case of lemongrass, it is 400kg of moisture-laden greenery that goes in and about 60 kg of dried product that comes out. The tea range now includes Peppermint, Spearmint, Lemon Mint, Chamomile and 3 blends of Ginger tea, all of which can be found in the local store, perched on shelves alongside much of the other regional produce.

Gaucamole Dip & Macadamia Bread

Guacamole would have to be a party favourite. I often add a dash of tabasco sauce if I want to spice it up a bit. Macadamia Bread can be easily made with pre-baked pizza bases or turkish bread.

Macadamia Bread

1 quantity pizza base

3 tbs. macadamia butter

1 clove garlic - crushed

macadamia oil

Guacamole Dip

2 avocados

1 clove garlic - crushed

1 tomato - diced

1 tbs. sour cream

1 tsp. lemon juice

salt and pepper

coriander sprigs

Preheat oven to 240°C. Roll out pizza base and place on a baking tray. Brush pizza base with macadamia oil. Combine crushed garlic and macadamia butter and spread over pizza base. Bake for approximately 10 minutes or until golden, less if base is pre-baked.

While the Macadamia Bread is baking, remove pulp from the avocado and mash roughly with a fork. In a bowl combine, avocados, garlic, sour cream and lemon juice. When well combined add diced tomato and season with salt and pepper. Garnish with coriander.

Serve with hot Macadamia Bread, corn chips, grizzini or vegetable sticks.

Journeying on, you will pass a number of pineapple plantations, rolling hillsides - a patchwork of silvery green pinstripes, so picturesque it is hard to imagine that these patches of pattern and colour are workable crops, not a just a man made creation to please the eye.

Pineapples are Queensland's most popular fruit and nothing is more symbolic of the Sunshine Coast than this queen of fruits. Crowned with its spiky plume of foliage, the regal yellow fruit has become synonymous with sunshine and the tropics. There are no gigantic structures here to landmark the fruit, just a splendid landscape, though the iconic Big Pinapple is not that far away - part of another food trail perhaps.

Macadamia nuts would have to be Queensland's most prosperous crop and there are numerous growers scattered throughout the Sunshine Coast, all of them producing top quality nuts and utilising them in a variety of ways. One of the larger plantations in the area and perhaps the most picturesque would have to be **Jondell Macadamia Plantation**. It is an amazing property that has an almost stately presence with its grand tree lined driveway that weaves its way to an equally grand homestead.

Jondell Macadamias have created their own niche market producing a range of flavoured nuts and their own brand of cold pressed Macadamia Nut Oil, which has the most fabulous flavour.

Avocados

Avocados are considered to be the world's most nutritious fruit. When in season they appear abundantly on roadside stalls throughout the coast.

Look out for:

- Fresh Avocados
- Avocado Oil
- Avocado Dips
- Avocado Mayonnaise
- Avocado & Cheese Spreads

Salmon on Hommus with Pesto & Macadamia Nuts

SALMON ON HOMMUS WITH PESTO & MACADAMIA NUTS

4 salmon fillet pieces

1 tbs. butter

3 tbs. Pesto

1 cup Hommus

½ cup macadamia nuts - toasted and coarsely crushed

snow pea shoots

basil

Heat butter in a frying pan. When hot place in salmon pieces and cook on each side approximately 3-5 minutes. Place a dollop of hommus on the centre of each plate and spread to a diameter of 10-12 cm. Place some snow pea shoots and basil sprigs on top of hommus.

When salmon is ready place on top of sprouts sprinkle with macadamia nuts and drizzle with Pesto. Serve.

Serves 4.

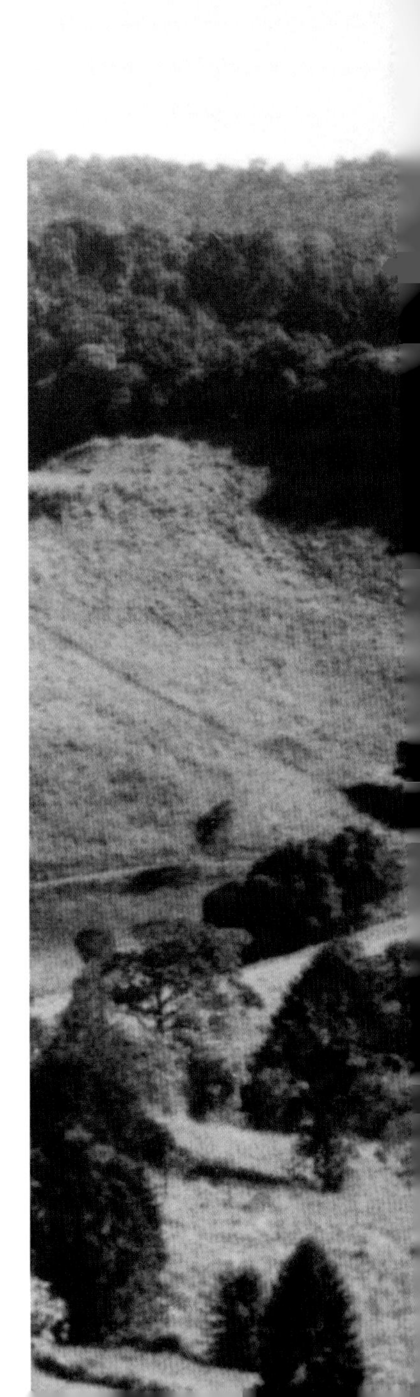

Native Bush Foods

There are a number of Bush Food producers emerging. Lloyds Fine Foods, Cedar Creek Farm and Galeru boasting the largest variety of individual products.

Look out for:

- Lemon Myrtle
- Desert Tomatoes
- Wattleseeds
- Davidson Plum jam
- Quandong relish
- Native fruit cake
- Rainforest Liqueurs
- Lilly Pilly berry conserves

Native to Australia, the macadamia tree is an intriguing specimen. Flowering macadamia trees have the most intense fragrance so it is quite a sensation to wander through a plantation during its prime flowering time. The nuts grow in clusters protected by a hard fleshly outer husk, which cracks open when the macadamia ripens to expose yet another shiny protective layer. They require effort to crack open, but once the nut has been conquered it is almost impossible to stop at one.

Although Jondell Plantation is not exactly open to the public, if you do a drive by, past the plantation in spring, with your windows open I am sure you will get to appreciate that amazing heady aroma of the macadamia blossoms, even if it is just for a fleeting moment.

Something I have always been passionate about is the native food industry. I don't think that we have yet truly discovered its enormous potential.

Native foods are our own indigenous foods, they are the foods that were here long before we ever were and have provided nourishment to the aboriginal communities for thousands of years. They are the fruits, nuts and seeds of native shrubs and trees - wattleseeds, sandpaper figs, lilly pilly berries, Davidson plums or in the instance of lemon myrtle, its leafy greenery. Each has a flavour quite unique resulting in some exceptional products emerging on the market.

Turkey Loaf

The loaf can be served hot, straight from the oven or served up in thin slices cold. Galeru make an unusual conserve out of the lillypilly berries, which compliments the turkey well.

1 single turkey breast (approx. 1.5 kg)

1 capsicum

5 slices prosciutto

1 small bunch of spinach

½ cup macadamia paste

2 tbs. macadamia oil

Galeru's Rainberry Conserve

Prepare the capsicum by cutting into quarters, de-seeding them and then placing them under a hot grill or over an open flame until the skin is scorched. Place capsicum pieces into a sealed plastic bag for a few minutes to allow them to sweat. The skin should now come off quite easily. Set aside. Rinse spinach leaves and set aside also.

Prepare the turkey by placing the turkey breast on a large chopping board and by using a meat tenderiser, even up the thickness of the turkey. By placing a sheet of plastic wrap on top of the turkey you can minimise damage to the flesh. This can also be done with a rolling pin. Try to urge the turkey breast into a rectangular shape approximately one inch thick.

Place a large sheet of plastic wrap onto the surface that you are to work on. Place the prosciutto slices in a row vertically along the centre of the plastic, roughly the same size and shape of the turkey. Then lay the turkey on top. Spread the macadamia paste evenly over the turkey, layer with spinach leaves and then the capsicum pieces. The turkey is now ready to roll up. Use the plastic to help keep everything in place pulling the plastic back as it is rolled in. Discard plastic.

When the turkey is rolled up, place it onto a lightly greased baking dish, seam side down. Use several skewers to secure the turkey in a firm roll while it is baking otherwise it will tend to want to unroll. Drizzle macadamia oil over the roll and then cover the tray with foil. Bake at 240°C for ½ an hour. Remove foil and bake for a further ½ an hour.

To serve hot, slice into inch thick pieces and serve with some Rainberry conserve.

Citrus

Oranges, mandarins and lemons are the Sunshine Coast's key citrus fruits, the kaffir lime joining them on a somewhat smaller scale.

Look out for:

- Kaffir lime marmalade
- Preserved lemons
- Brandied mandarins
- Lime pickle
- Limoncello liqueur
- Lemon butter
- Orange pepper

Cooroy seems to be the home to quite a few native food specialists, **Lloyds Fine Foods** and **Galeru**, in particular, producing top quality products that follow a native theme while still remaining mainstream enough to encourage people to try and be pleasantly surprised by what they have tasted. **Lloyds Fine Foods**, originally known as The Gourmet Gum, utilise bunya nuts and macadamia nuts in most of their pestos. They produce the most delicious Bush Tomato Relish and have developed recipes that include some of the more unusual native fruits such as quandongs and native tamarind. As is the case with most producers, they utilise the fruits from their own resources resulting in quite a burgeoning little industry. They all have different methods, some choosing to regenerate existing native land and planting to complement what is already in place while others choose to take a more crop style approach, either way I think it is fabulous that theses native resources are no longer being ignored.

Galeru have taken a slightly different approach to the bush food concept. Having planted an extensive range of native shrubs and trees on their property, they were looking to explore ways of utilising the various berries and fruits that were harvested. Among these are lilly pilly berries and the fruit of the Davidson plum, which they have developed into decadent baked products and sauces. The berries have an aromatic spiciness which when in combination with the luscious richness of cakes such as their Chocolate Rainberry Truffle Cake are absolutely divine.

Cooroy is a great little detour in itself, the Old Butter Factory site hosting many exhibitions, showcasing the local talent, artisan or otherwise.

Another wonderful Food Trail to follow is the route that takes you through Kenilworth. From there you could head north and end up at Imbil, or south through the Obi Obi Range and back through Maleny. There are wineries, cheese makers, a deer farm and a multitude of hidden pockets that have various boutique crops that probably won't be as apparent on your first trip.

One of these pockets hides **Cedar Creek Farm**, although they aren't open to the public they do have stalls at many of the local Farmers Markets and their products are well worth looking out for. They are hidden amongst a glorious patch of rainforest, which canopies a creek that has carved a passage through the property. A rickety, moss covered timber footbridge spans a dreamy waterhole which, incidentally has provided the patient with quite a few fish dinners. The property itself has been the home to numerous experimental crops, but lemongrass, mandarins and kaffir limes are now the main crops, providing the base to their range of condiments and preserves. Infused oils such as their Kaffir Lime Oil are sensational and their Preserved Lemons are incredibly more-ish. Cedar Creek Farm also favours using native products and are currently experimenting with native spices such as pepperberries and lemon thyme with some interesting results.

En route to Kenilworth is **Belli Bamboo Parkland** a property of extraordinary proportions. Thirteen years of collecting, cataloguing and cultivating bamboo now sees the property boast over 250 varieties of bamboo. Bamboo takes many shapes and forms, which is apparent from the moment one enters the property via a pretty magical bamboo lined drive.

Bamboo

Edible bamboo shoots are incredibly nutritious. They can be eaten fresh or boiled. They make a refreshing addition to stir fries and Asian style dishes.

Look out for:

- Fresh bamboo shoots
- Bamboos shoots on restaurant menus

Textures abound, from smooth poles that shoot like multihued pillars from the earth to the velvet husks that peel from the shoot itself. Bamboo groves have a spiritual quality, an ethereal silence that evoke a feeling of peace and serenity, apparent to all that have visited the property's running bamboo grove. Belli Parkland – Bamboo Australia is open to the public for sales. If you tour the grounds, the running bamboo grove is a visual delight, cleverly opening up into an arena. Intended for functions it has a surreal ambience that would be hard to find elsewhere.

From a gourmet perspective, edible bamboo shoots are incredibly nutritious, but like so many 'new' crops, relatively unexplored by Australian consumers. Bamboo shoots can be eaten fresh or boiled and make a refreshing addition to stir fries and Asian style dishes.

Dairy Products

Wonderfully, the Sunshine Coast has gained a number of independent dairies that have managed to prise their products on to supermarket and convenience store shelves as well as supplying other industry producers. Noosa Eumundi Dairies and Maleny Dairies are the main suppliers but there are smaller dairies also.

Look out for:

- Fresh milk & cream
- Yoghurts & custards
- Hand crafted ice creams

Kenilworth is the heart and soul of the Mary Valley, a beautifully lush and fertile region of the Sunshine Coast's hinterland. It is a picturesque little town, full of charm and character. It is rich in dairy history and community mindedness, a history that has managed to set this town apart from the rest.

Kenilworth Country Foods or the **Kenilworth Cheese Factory** as it is more commonly known, tells a story that is hard to beat. The factory, opened in 1952, was originally a Kraft Cheese factory, which when faced with closure saw the local community band together to protect their jobs and local industry. Six workers at the factory and an investor took a huge leap of faith and purchased the factory going

Grilled Duck with Rosti, Lychee Salsa & Rosella Pepperberry Sauce

4 single duck breasts

¼ cup Rosella Cordial

1 tbs. olive oil

1 tbs. ground pepperberries

1 clove crushed garlic

salt and pepper – pinch

1 bunch watercress

Lychee salsa:

½ kg lychees

1 clove garlic – crushed

1 shallot – sliced

1cm piece of ginger

1 tbs. olive oil

1 tbs. lemon juice

salt and pepper to taste

Potato and Ricotta Rosti:

2 medium potatoes

½ cup Ricotta – drained

pinch of salt

½ lemon – juice only

oil for frying

Mix together ingredients for the marinade and pour over the duck breasts. Coat each side and allow to sit for an hour in the refrigerator. Prepare rostis and salsa. To prepare the salsa, peel and seed lychees and cut in half. Add remaining lychee salsa ingredients, combine and set aside. For the rosti, roughly grate the potatoes and add the lemon juice to it immediately. Combine well. The lemon juice stops the potatoes from turning grey. Add the ricotta and combine. Add salt. Set aside.

When dinner is ready to be served, you must move quickly and efficiently, so having everything ready in advance helps. Heat the grill. Place the duck skin side down on the grilling tray. Grill for approximately 7 minutes, a little bit longer if they are particularly large. Turn them over and grill for a few more minutes until skin is a beautiful golden brown. At this point don't take your eye off the duck for a moment because it can go from golden brown to burnt in just seconds. Turn off the grill and allow the duck to rest for approximately 10 minutes.

While the duck is resting, divide the watercress between each plate and place a scoop of the lychee salsa on each plate as well. When the 10 minutes is almost up, prepare the rosti and the sauce. Heat the marinade in a small saucepan bring to the boil and reduce for 1 minute, then turn off. Heat the olive oil in a frying pan. Drain off and squeeze excess liquid from the potato and ricotta mixture. Divide into four portions and form each portion into a flat pattie. When the oil is sizzling hot, place the patties in the pan and fry a few minutes on each side until golden brown and beginning to crisp.

To serve, place a rosti on the bed of watercress and place the sliced duck breast on top. Drizzle with rosella and pepperberry sauce. Serve immediately. Serves 4.

Goat
Goats Milk
Rennet
Basil – Garlic

on to make boutique style cheeses in the tradition of its predecessors. The factory continues to make award winning cheeses including the original Malling Red and Malling Roma, both sharp tasting textural cheeses that have been made since the factory's inception in 1990. Over the years the factory has developed quite a repertoire of cheeses from the smooth delicate flavours of the Kalaas to stronger ones such as the Vintage or the more uniquely flavoured cheeses such as the Spiced Gouda and the Smoked Sun-dried Tomato Cheese. They also make a variety of yoghurts and custards and a creamy ricotta.

Going to the factory allows you to sample all the different styles available. So drop in, stock up on your favourite cheeses, grab a loaf of bread in town and pull in at one of the many spectacular picnic stops in the area, not forgetting to veer past the **Kenilworth Bluff Winery** for a bottle of red on the way.

For those that would like to get a little more involved with the cheese making process, an absolutely delightful experience is that of **Coolabine Farmstead Goat Cheeses**. Coolabine Farmstead is a short drive from Kenilworth and is open to the public for cheese tastings and farm visits several days a week and for those who wish to further the experience there are Cheese making workshops. There is something about wandering about the farm and mingling with the goats and their animated young that make this experience thoroughly unforgettable. The goats are like part of an extended family so they are curious and affectionate with visitors and manageable when it comes to milking. The cheese making process itself is fascinating and through the workshops there is the opportunity to really understand the popularity of these artisan cheese industries. The cheeses are

Cheese

Cheese making has quite a few stellar performers on the Sunshine Coast. While Gympie Cheese and Coolabine Farmstead's focus is on handcrafted goats cheese, Maleny Cheese and Kenilworth Country Foods are producing some fabulous everyday cheeses.

Look out for:

- Fresh Fetta
- Fetta in oil
- Chevre
- Camembert & Brie
- Sun-dried tomato & wasabi flavoured cheeses

Roast Spatchcocks with Preserved Lemon & Mushroom Risotto

This has become one of my favourite dishes, the preserved lemon a fabulous complement to the earthy tones of the mushroom risotto and the spatchcock. A good quality stock is the key to a good risotto. Try making your own stock, but if time doesn't permit there are some good quality pre-prepared stocks available in the supermarket. Grilled duck is a great alternative to spatchcock and even chicken shines with this dish.

2 spatchcocks – halved

1 tbs. Curry Powder

1 tbs. butter

2 cups arborio rice

1 onion – finely chopped

4 tbs. butter

200g Swiss Brown Mushrooms - sliced

6 cups chicken stock

1 cup white wine

1 preserved lemon – rinsed, flesh removed & finely chopped

½ cup parsley - chopped

Preheat oven to 240°C. Place spatchcock halves on a baking tray, skin side up. Melt butter and brush over spatchcock liberally. Sprinkle the curry powder over spatchcock and rub into flesh. Place in the oven and roast at 240°C for approximately 30 minutes, basting it with its own juices after about 10 minutes. The spatchcock is ready when it is pierced with a skewer and juices run clear.

While the spatchcock is roasting begin the risotto. In a large saucepan melt the extra butter and sauté the onions for one minute. Add the rice and stir until the rice has absorbed all of the butter. Do not allow the rice to stick to the saucepan. Add the wine stirring constantly. When the rice has absorbed the wine, add the stock one ladle at a time, each time stirring continuously until the liquid has been entirely absorbed before adding another ladle. This is an arduous process but it is the secret to a perfect risotto. When you are three-quarters of the way through the stock, add the mushrooms and then continue adding the stock. The risotto and spatchcock should be ready at roughly the same time. If the spatchcock is ready first, cover with foil and sit in the warm oven until the risotto is complete. When the last ladle of liquid is added, the risotto should be ready and beautifully moist, not gluggy. Stir in preserved lemons and parsley and serve immediately.

Serve ladled into a deep dish with half a spatchcock on each dish. Garnish with a little parsley and a wedge of lemon. Serves 4.

uniquely delicious and available for purchase from the farm although it is wise to ring prior to visiting to avoid disappointment.

I have always enjoyed visiting Yandina, the first visit being prompted by a certain sense of obligation to make the expected pilgrimage to **The Ginger Factory**, but like a magnet, Yandina draws the unsuspecting foodie back time and time again. Saturday markets, the Wappa Ponds Herb Farm, Fairhill Nursery – a place to stock up on your native bush food plants – and a place of beauty, **The Spirit House**.

It would sound slightly more magical if I could say that The Spirit House is shrouded in mist and only the true at heart find their way through the entrance, but this restaurant is open to all and over the years its sensational menus and cooking school have become something of a gourmet institution.

A bamboo-lined path leads you through a lush tropical garden, bamboo torches and fairy lights lining the pathway at night. The Thai style open building, a spirit house, is the restaurant. Spirit houses are common throughout Thailand. Here offerings of fruit, flowers and food are placed so that peace and prosperity is brought to the household by its guardian spirits. The ambience of the gardens and the glow of the lanterns that flicker their reflections in the papyrus-lined pond that the Spirit House sits beside, gives an ethereal quality and a sense of peace and well-being. The music mingles with the aromatic scent of ginger and exotic spices that dance their way from the kitchen to the diner to provide a setting that could not be a more

Ginger

Ginger would have to be one of the most versatile ingredients. Buderim Ginger seems to have covered all bases, producing marmalades, sauces, pickled ginger, crushed and grated ginger, ginger in syrup etc.

Look out for:

* Pickled Ginger
* Ginger cordial
* Ginger Gummi Bears
* Ginger Beer – alcoholic & non alcoholic
* Ginger tea
* Ginger ice cream

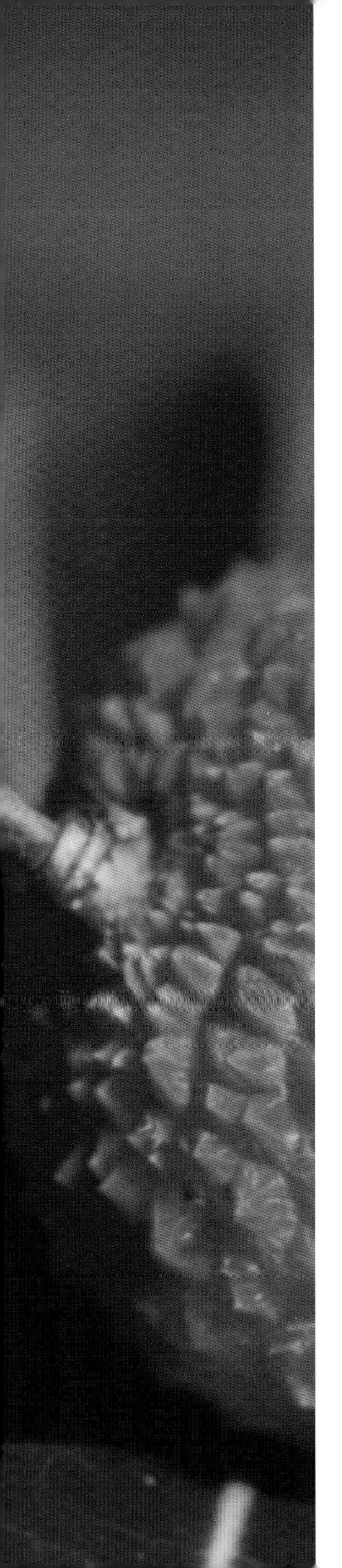

Lychee & Chilli Ice Cream

This is quite a unique, refreshing ice cream. The chilli compliments the lychee well. A variation of the recipe is to substitute finely chopped Kaffir lime leaves in place of the chilli.

300g lychees – shells and seeds removed
250g castor sugar
250g water
400ml coconut cream
1 chilli – deseeded and finely chopped
200g yoghurt
juice of 1 lime
1 tbs. Contreau

Place lychees in a food processor and process until pureed. Heat the water and sugar and bring to the boil. Simmer for 5 minutes and then remove from heat and allow to cool.

In a bowl, combine the sugar syrup, lychee puree, coconut cream, chilli, yoghurt, lime juice and Contreau. Pour mixture into an ice cream machine following the instructions of the manufacturer.

Serve with fresh lychees soaked in Contreau.

breath taking introduction to fine food.

The Spirit House is a sanctuary to Asian style cuisine, offering a cooking school with a multitude of classes as well their own selection of aromatic curry pastes and take home meals. The Spirit House chefs have always been an encouraging support to local growers, sourcing their produce, wherever possible, locally, adding to their own assortment of Asian herbs and leafy greens that are grown hydroponically on the property and supplied to other local restaurants as well. The Spirit House is a must for any visitor to the Sunshine Coast.

Throughout the Yandina region you will see acres of ginger growing, depending of course on what the season is at the time. Autumn is when it is at its most apparent, just before the bulk of the harvest. Ginger has long been known for its medicinal properties and has gained a reputation in Indian myth for awakening one's 'inner fire', which nurtures divinity and creativity. Ginger originated in China and India and through various diversions from the spice route made its way to Australian shores in the early 1900's. The Queensland conditions were ideal for growing this ambrosial rhizome and plantations were quickly established throughout Queensland. **Buderim Ginger** now produces a large percentage of the world's processed ginger, and the Sunshine Coast is where it nearly all grows. The Ginger Factory began as a cooperative in Buderim by five local farmers. By the late 1970's it had outgrown its Buderim premises and relocated to Yandina in 1979, to what has grown into one of Queensland's most popular tourist attractions. Buderim Ginger produces a fabulous range of ginger products making it hard to pick a favourite, but if you are a ginger fan there will be definitely something to take home.

Lychees

Lychees are a sweet sensual fruit, very rich in Vitamin C. They are available throughout summer and are delicious stuffed with cheese or pestos, or even better, chocolate!

Look out for:

- Fresh, fresh, fresh lychees!
- Liquored lychees
- Lychee sorbet
- Lychee wine
- Lychee salsas

HIGH FIBRE
HIGH PROTEIN
PRESERVATIVE FREE
CERTIFIED ORGANIC
TEMPEH
Curry Soy

Prawn, Tempeh & Macadamia Lemongrass Sticks

PRAWN, TEMPEH & MACADAMIA LEMONGRASS STICKS

10 sticks of lemongrass
250g green prawns
125g fresh tempeh - roughly cubed
½ cup macadamias
1 clove garlic - crushed
1 tsp. lemongrass - finely chopped, extra
1 tsp. ginger - finely chopped
½ chilli - finely chopped - optional
½ an egg white
salt and pepper
plain flour
macadamia oil

Trim lemongrass off the excess foliage and cut each to a length of about 20cm. Set aside. Heat a little macadamia oil in a frying pan and lightly fry tempeh until golden brown. Set tempeh aside on some absorbent paper until cool.

Lightly toast macadamias in the same pan and set aside to cool also. Shell and roughly chop prawns and place in a food processor with the macadamias, tempeh, ginger, garlic, extra lemongrass, chilli, egg white, salt and pepper. Process until mixture is an even consistency. Divide mixture into 10 portions and with a little flour on your hands, mould each portion around a lemongrass stick.

When ready to serve, grill on the BBQ or in a frying pan with a little oil, turning until each side is golden brown. Serve with wedges of lemon and a variety of dipping sauces.

Tempeh

Tempeh is a nutritious Soya bean product. It is best bought fresh from the markets or health food stores, which can then be frozen for use at a later date.

Look out for:

- Marinated tempeh
- Tempeh burgers
- Tempeh 'soyalamis'

Like ginger there are many industries on the coast that have in some way been inspired by the similarities in climate with the Asian tropics. Lychees, mangos and a multitude of other experimental fruits have blossomed into booming Queensland industries. Lychees in particular are fast becoming one of Australia's leading exports rivalling countries such as Madagascar and Mauritius.

Lychees originally from China, are an intriguing little fruit. A bright red, armour like shell protects its delicate, pearly flesh, which is slightly musky, almost peppery in taste. Held in high regard by the Chinese, who have cultivated lychees for thousands of years, they are considered to be a symbol of romance and good fortune. Fresh, the lychee offers unlimited possibilities – lychees salsas, stuffing fresh whole lychees with sweet or savoury fillings or as an interesting addition to curries and salads. **Flower Fruits** is a major grower, but there are quite few orchards, located throughout the Sunshine Coast some of them quite happy to sell the fruit from roadside stalls when they are in season.

One discovery is the tempeh produced by **Mighty Bean Soy Foods**. One tends to think of of tempeh and soy related products in terms of health food and although they are, one should not be inhibited from actually trying a lot of these foods from a gourmet perspective. Since first discovering Mighty Bean's tempeh, It is hard to find a tempeh that equals it in quality or taste. It is a big call, but this tempeh is

Fried Tempeh

Fried tempeh drizzled with a little pesto is just magic and makes a perfect light summer lunch or dinner. Search out the locally made tempeh, by Mighty Bean if you can because it is like no other tempeh that I have ever tasted. There is quite a range of savoury pestos made by local producers to choose from, some producers incorporating native ingredients such as bunya nuts or macadamia nuts for a slight twist.

1 x 500g packet tempeh

rocket leaves

oil for frying

1 mango - diced

1 inch grated ginger

1 clove garlic - crushed

1 spring onion - chopped

½ chilli - chopped finely

1 tsp, lime juice

1 tsp. olive oil

¼ tsp. palm sugar

salt and pepper

½ cup Pesto

To make the salsa, place the diced mango, ginger, garlic, spring onion and chilli in a small bowl. Sprinkle the ingredients with salt, pepper, palm sugar, limejuice and olive oil and combine together very gently. Set aside.

Cut tempeh into slices approximately 12mm thick. Put enough oil in a frying pan to coat the bottom of the pan. Heat the pan gently until the oil is hot and place in the tempeh slices. Fry tempeh on each side until golden brown. When cooked, allow the tempeh to drain on absorbent paper for a moment.

To serve, place a large spoonful of the mango salsa on each plate. Arrange a few rocket leaves next to the salsa and place a few slices of the tempeh on top of them. Drizzle with pesto and garnish with basil. Serve while hot with a wedge of lime.

Serves 2-4.

seriously good, not just healthy!

Tempeh is a product made through the fermentation and culturing process of the soya bean. It is a three-day process, involving soaking the beans, dehulling them and then introducing a spore, Rhizopus Oligosporus, which in turn creates the tempeh after a short incubation period. Tempeh has a soft, grainy texture, is slightly nutty and an amazingly tasty and versatile food. Soya beans are nutrient rich and considered to have health giving properties. This aside it can be served a dozen different ways – grilled, deep fried, marinated, in sushi, on the BBQ, in curries or stews. Mighty Bean's tempeh is made on site in a small processing facility just out of Yandina. It is sold fresh at the Eumundi Markets on Saturday mornings. Also available are tempeh sausages, burgers as well as a range of marinated tempeh products.

Another Yandina local is the **Wappa Ponds Herb Farm.** The farm itself is picturesquely nestled into the hills, elevated just enough to appreciate the tranquil beauty of the Wappa Dam. In the hydroponic gardens, natural eco systems are maintained and encouraged; thereby creating a wonderful cohesion between what has been put in place and what previously existed. The list of herbs on offer is constantly growing and it would probably suffice to say that the herb farm grows most of the more common culinary herbs and many of the lesser-known herbs, all of which are available for purchase. The Herb Farm is open to the public. Private tours and farm walks are by appointment and are an enjoyable and informative way of learning more about herbs and how to grow them.

Herbs

There are a number of herb growers open to the public stocking endless lists of rare and exotic edible, nutritional and medicinal herbs as well as the more traditional herbs. Some herb farms offer tours, or give, growing hints and share their own extensive knowledge.

Look out for:

- Traditional herbs
- Asian herbs and leafy greens
- Vanilla vines
- Spice trees
- Native herbs
- Wasabi

OREGANO
ROCKET
Wappa Ponds
Farm 5446 8425
TARRAGON
LEMON

Mushrooms

Swiss Browns and White Agaricus are the most common mushrooms grown in this region although a few growers are currently experimenting with some of the more exotic varieties.

Look out for:

- Marinated mushrooms
- Mushroom and walnut pâté
- Mushroom, coriander and almond pesto
- Mushroom tapenade

What leads a person into the industries that they so passionately pursue? Whether it be herbs, or cheeses, or some exotic fruit that nobody has ever heard of it is intriguing. When it comes to the choice of mushroom growing, I mean mushrooms, they hardly seem inspiring. What is it? An opaque little fungus, thriving in dark, dank space, but the growing process is quite extraordinary. Tiny round buttons miraculously appear, multiplying in size and number at an incredible rate. In a growing room I once visited, the light filtered in through an open door and suddenly bags that were filled with mushrooms at various stages of growth started to look like glistening fairy rings and I begin to understand a little bit of that passion.

There are a few growers on the coast. One of the more serious growers of Swiss Brown Mushrooms is **DJ's Mushrooms**, whose collaboration with Cedar Creek Farm has resulted in some pretty wonderful mushroom based products. The Mushroom Paté and Mushroom Tapenade are exceptional. Once again, these products can usually be found at the Eumundi Markets, but most of the local markets seem to have at least one of the local mushroom growers there with a stall.

Mt Coolum

GRILLED MANGOS

Grilled Mangos

Mangos need little in the way of enhancing. They have an abundance of natural flavour that is unparalleled by any other fruit. I don't think the mango is a fruit that responds well to cooking, so quickly grilling them in this manner allows the mango to retain its shape and flavour. Use an Almond liqueur such as Amaretto or even better, Moonshine Valley's Almondo, a local version.

mangos
brown sugar
Almond Liqueur
almonds - flaked
cream – lightly beaten

Cut a half of each side of the mango and score the mango flesh with a sharp knife being careful not to pierce the skin. Carefully pour some Almond Liqueur into the scores of the mango flesh. Spread some whipped cream thickly on to the surface of the mango. Sprinkle the top with almond flakes and brown sugar and grill under a hot grill for a few minutes until almonds start to brown.

To serve, turn flesh inside out and pour over any juices that escaped while grilling.

Mangos

The mango has an aromatic soft flesh, which has made this luscious fruit a tropical favourite. Mangos are extremely rich in Vitamin C as well as Vitamin A. If you are not lucky enough to have a mango tree in your garden I am sure you won't have any trouble at all finding them when they are in season.

Look out for:

- Mango jams
- Mango syrups and drinks
- Mango ice cream
- Mango chutneys
- Mango Daiquiris!

Mooloolaba

When it comes to coastal crowns, Noosa once reigned supreme and to many it still does, but over the years Mooloolaba has undergone an extraordinary metamorphosis seeing it slowly sidle its way up the ranks to demand its share of attention and gourmet praise. Sleek trendy cafés and restaurants line the esplanade like glamorous toy solders, dozens of tables speckling the sidewalk with waiters strutting the culinary catwalk from the kitchen to the diner, delivering mouth-watering morsels. Mooloolaba's streetscape faces the sea so diners can sit and absorb the salty sea breezes, sunshine and spectacular views. Like Noosa, the focus is on local produce. Seafood being a natural feature on most menus due to the close convenience of a very prosperous fishing industry.

Mooloolaba has long been the major port of call for commercial fishing boats and trawlers. Prawns, crabs and tuna, along with a multitude of other seafaring edibles are brought into its bustling wharfs to be distributed throughout Australia and overseas.

The Mooloolaba Spit itself has become a seafood mecca with a number of wholesalers offering fresh quality seafood direct to the public – it's about as close as you can get to buying it directly from the fishermen themselves. Some wholesalers also offer dine in or take-away meals. There probably isn't a person alive that wouldn't appreciate the simplicity of eating fish and chips on the beach while watching yachts sailing in after a day out at sea. Although Mooloolaba and Maroochydore are certainly up there with the best as far as fine dining and fresh produce is concerned, the atmosphere on a whole is welcomingly, a little more laid back.

The coastal stretch from Mooloolaba to Maroochydore remains a public domain, very much pedestrian orientated with walking and bicycle paths joining the two ends of a very pretty strip of coastline. Maroochydore has undergone its own glossy transformation of late with the development of the old CBD into a very hip riverside promenade, allowing diners to gaze out dreamily over the river. The river itself is laden with fish, anglers flocking to popular spots like Chambers Island or Cod Hole to try their luck. All in all, Mooloolaba and Maroochydore have as much to offer as any seaside town with the

TUNA SASHIMI

Tuna Sashimi

When purchasing tuna for sashimi always ask for 'sashimi grade' tuna - it must be absolutely fresh as it is eaten raw. Try this recipe using an infused oil such as a chilli oil or kaffir lime oil for a spectacular result.

500g sashimi grade tuna

1 tbs. - herb of your choice - dill, coriander, chives - finely chopped

pepper - freshly ground

1 tbs. salmon roe and/or finger lime pulp

2 tbs. olive oil

1 tbs. wasabi paste

pickled ginger

Freeze tuna for an hour prior to serving - this will allow the tuna to be sliced finer.

Prepare the platter by scantily spreading the wasabi paste over the area where the tuna is to be placed. Slice the tuna as finely as possible and arrange on the platter - choose a large platter so that the slices do not overlap. Sprinkle tuna with herbs, salmon and/or fingerlime pulp and pepper. Drizzle with oil. Garnish with pickled ginger - the more talented among us may construct a rose from the ginger slices, but otherwise a neat little pile will be fine.

Serve with little dipping bowls of lemon juice and soy sauce.

Ocean Wanderer
fish
on parkyn
bistro
bar
takeaway
See

HKG
1088
A 60

5959
9000

WHITING IN CHILLI BEER BATTER

Whiting in Chilli Beer Batter

Whiting have a lovely flavour and the chilli beer batter adds that extra little bit of zing. This recipe also suits using ginger beer, as an alternative to the Chilli Beer.

6 small whiting fillets

1 cup self-raising flour

1 cup Chilli Beer

1 egg yolk

1 pinch of salt

*1 tbs. black sesame seeds**

extra flour for dusting

oil for deep-frying

Place the self-raising flour and salt in a bowl and make a well in the centre. Slowly pour the beer in the well, small portions at a time, stirring the flour into the mixture slowly from the sides. When all of the beer has been added, then beat in the egg and add the sesame seeds. Place in the refrigerator for an hour and allow the batter to rest.

When the fish is ready to be served, heat enough oil in which to immerse the fish. Dust each fillet with a little flour on each side and then coat with batter. Allow excess batter to run off and then place in hot oil. Fry for a few minutes on each side, until each side is golden. Remove from oil and allow to drain briefly on absorbent paper.

Serve immediately with some salad, mayonnaise or tartare sauce and wedges of lemon or lime….. oh yeah….and a Chilli Beer as well!

* black sesame seeds are available from Asian grocery stores or just use the regular sesame seeds.

added bonus of a produce rich hinterland just a short drive away.

Seafood

Between Noosa and Mooloolaba there are dozens of fresh seafood outlets with widening selections of fresh fish and crustaceans.

Look out for:

- Tuna
- Mahi Mahi
- Locally farmed barramundi
- Red claw crayfish
- Scallops
- Oysters
- Moreton Bay bugs

With all this wonderful food, alcoholic beverages are not to be overlooked. Whether it is wine, beer or fine liqueurs, all are developing a strong industry presence on the Sunshine Coast.

A popular pastime, beer and the consumption of it are very much a part of the Australian psyche whether we care to admit it or not. At this very moment, men in sheds and garages all over Australia are labouring to create a 'top-drop' – thirst quenching ales, brewed to perfection with home brew kits that were probably given to them for Fathers Day or Christmas.

What constitutes a perfect draught is up for debate, everyone's taste is different, dictating why there are so many beers to choose from. **The Sunshine Coast Brewery** in Kunda Park offers over 20 different flavours of beer, all brewed on the premises and as many as 12 of these are offered on tap in the bar.

The brewery sticks to more traditional brewing techniques to produce their beers. The process takes about 3-4 weeks and a little more complicated than your standard home brew kit, but in short the grain is cracked and mashed and once the malt extract and hops has been added, it is chilled and then fermented. When the brew has matured, it is filtered, carbonated and bottled or sold on tap.

The brewery has also produced some unusual tropically inspired beer in the way of a Chilli Beer and more recently, an alcoholic Ginger

Beer made using local ginger. If traditional beers are more your style, there are plenty of those also. The Sunshine Coast Brewery is open to the public daily, offering weary travellers and thirsty locals a reprieve as well as the chance to sample one or more of their quality brews.

Only a short drive further will bring you to Forest Glen's **Moonshine Valley Winery**. Originally the home to fruit wines, the winery has shifted its focus to grape wines and quality handmade liqueurs. Although there is no vineyard on site, the winery is producing top quality wines with grapes purchased from committed growers that have mastered Queensland's climatic conditions.

Reading wine labels is a bit like reading poetry. Clusters of evocative prose create a sensual imagery that turns a voluminous liquid into the forms of voluptuous women or floral bouquets amidst wild brambles and oaken forests.... full bodied, earthy, luxurious, berry or citrus notes, the words just flow, as does the wine. Moonshine Valley's labels are no different, 'luscious handcrafted jewels' such as their Chardonnay, Verdelho and Shiraz are crafted to please a variety of palates.

The winery also produces some 'to die for' liqueurs and fortified wines. There is a luxurious coffee liqueur, Espresso, made out of ground Queensland coffee beans as well as a fabulous Old Buderim Ginger Port. One favourite is La dolca vita, a white port made in the style of the Swiss Ice Wines, where the grapes are picked frozen from the vines after the first frost. Recreating this wine was a challenge due to our somewhat more exhausting climate but after a few improvisations they have managed to produce a sweet fortified wine that is everything and more than the label says. Moonshine

Beer

Traditional natural brewing methods are proving a favourable alternative to mass production beer. Smaller boutique style breweries create a string of successful brews.

Look out for:

- Sunshine Coast Bitter
- Robinson's Stout
- Chilli Beer
- Ginger Kegs

original
beer
Wild Raspber
appeal of the
Indulgence. The
forth into the sunshine
fringing the creeks

ROBINSON'S
CHILLI BEER
330mL
375ml
Moonshine Valley
"La dolce vita"
WINEMAKER
Barossa Valley
AMERICAN OAK
MEDIUM GRAIN
M PLUS / T.T
2002

Liqueurs

Do not underestimate the quality of our local liqueurs. There are some exquisite flavours available at select wineries and specialty liquor stores.

Look out for:

- La dolce vita
- Old Buderim Ginger Port
- Espresso Coffee liqueur
- Limoncello
- Mango liqueur
- Rainforest Liqueurs
- Fruit wines

Valley Winery is open to the public for tastings most days and is a great opportunity to get to know the wines and a little more about the industry in general.

Stop in at the **Organic Oasis** and the **Organic Butcher Shop** at Forest Glen while you are there. Both are huge local industry supporters, the butcher shop stocking poultry from **Bendele Farm** (Kilkivan) and **Dakota Free Range Turkeys** (Kin Kin) while the Organic Oasis stocks nearly all of the local organic produce.

Honeybees would have to be one of the natural wonders of the insect world. These nectar-seeking maestros diligently compose a symphony of waxen cells to create mazes filled with their own natural alchemy of liquid gold. Beekeeping is an ancient art, originating from an era of myth and fable, an art that has not diminished with evolving traditions and science.

Gods of mythical worlds delighted in ambrosial honey based nectars and delicate sweetmeats. They employed the use of honey and the other treasures from the hive to create life-preserving balms. The medicinal use of honey is currently gaining momentum as scientists and medical practitioners are rediscovering the healing properties and medicinal qualities of this syrupy substance.

Espresso Cream Shots

If you have a penchant for coffee liqueurs you will find it very hard to overlook Moonshine Valley's Espresso Liqueur. It is a luxurious liqueur with the rich full flavour of the freshly ground coffee beans with which the liqueur is infused. The coffee itself is grown in Northern Queensland. The liqueur can be used as a flavouring for rich desserts or in one of the many cocktails in which coffee liqueurs are used, but there is probably nothing better than one of these shots with cream.

Moonshine Valley's Espresso Coffee Liqueur
cream
Coffee beans

Pour the Espresso Coffee Liqueur ¾ of the way up a small shot glass. Fill to the top with cream and plonk a coffee bean on top.

Serve.

Lemon Myrtle Snapper in Banana Leaves with Lemon Coconut Sauce

This is a great way to do a whole fish. The banana leaf helps the fish to retain its moisture while the fish is infused with the flavours of lemon myrtle and ginger. I prefer to use fresh lemon myrtle leaves if they are available, but if not just use the dried variety – lemongrass is a good substitute as well. You can use one large snapper, but it is just as easy to make single portions using 4 plate size snappers, allowing them to be served up individually rather than having to parcel out portions.

1.5 kg whole snapper – scaled and cleaned

5 spring onions

2 inch piece of grated ginger

6 lemon myrtle leaves – fresh

2 tbs. butter

1 tbs. olive oil

1 clove garlic - crushed

1 cup coconut cream

½ tsp. lemon myrtle – ground

1 tsp. lemon juice

1 tbs. fingerlime pulp

salt and pepper

1 large banana leaf

lemongrass stalks or raffia for tying

Preheat oven to 200˚C. Prepare the banana leaf by blanching briefly in a sink full of hot water. Set aside. Score the flesh of the fish 3 or 4 times on each side and rub salt into the gashes and the inside cavity of the fish. Chop 4 of the spring onions into inch (25mm) long pieces. Lay the banana leaf on a baking tray. Scatter 1/3 of the spring onions onto the centre of the banana leaf along with 1/3 of the ginger slices and 3 lemon myrtle leaves. Place the fish on the leaf. Fill the cavity of the fish with another 1/3 of the spring onions, ginger and lemon myrtle and scatter the remaining 1/3 on top of the fish. Dollop knobs of butter all over the fish. Fold the two ends of the banana leaf over the fish and then fold up the sides and secure into place with two or three blades of lemongrass. Bake fish at 200˚C for 45 minutes. While the fish is baking start preparing the sauce. Heat the oil in a small saucepan. Finely chop the last spring onion and place it and the garlic in the oil. Sauté for 1 minute over gentle heat. Add the dried lemon myrtle and pour in the coconut cream. Gently simmer, stirring occasionally, until the coconut cream has reduced enough to thicken the sauce's consistency. Add the lemon juice and simmer for another 2-3 minutes. The sauce can then be set aside and reheated when the fish is ready to be served.

When the fish is cooked, reheat the sauce, add the fingerlime pulp and heat through. Serve hot with wedges of lemon and steamed baby potatoes.

There are many different flavours of honey depending on the flowering fauna closest to the hive. Common to find on the Sunshine Coast is Macadamia, Rainforest and Orange Blossom, although there are many floral mixtures, all deliciously sweet.

Australis Bee Products retails honey and honeycomb products to a large sector of the local restaurant industry and delicatessens They take special care that the honey is kept at its purest. It seems there has also been an element of experimentation with the honey from the native bee, which hopefully we will see a lot more.

It is not hard to find local honey, most of the local markets have at least one honey vendor and roadside stalls prove fairly fruitful too. Another stop off is the **Superbee Honey Factory** at Tanawha. The factory has an enormous selection of honey and honey based products. It is also a supporter of other local industries. The honey-tasting bar allows visitors to sample the different floral flavours - you will be astounded at the differences. As for honey itself, research has shown that it contains many anti-bacterial agents, vitamins and minerals. It has long been used as a soother for sore throats and to fight the common cold and if you need any more excuses to hunt for the freshest and the purest, it tastes great too.

Woombye is an incredibly fertile region. As the home to one of the Sunshine Coast's iconic tourist attractions, the **Big Pineapple,** Woombye is no stranger to the region's fruit growing history. The area has seen the growth of passionfruit, custard apples, pineapples and acres upon acres of avocados. Figs are the most recent addition

Honey

Native flora, citrus blossoms and macadamia blossoms all impart their own delicate flavours to the honey that bees diligently produce. Markets, fruit and vegetable stores, delicatessens and health food stores all stock local honey, raw and in the comb.

Look out for:

- Macadamia honey
- Comb honey
- Native bee honey
- Creamed honey
- Honey based nougat

MACADAMIA NUT BAKLAVA

Macadamia Nut Baklava

Baklava is a Middle Eastern pastry, usually made with pistachio nuts or almonds. This recipe uses macadamia nuts, which can easily be substituted with either of the former. There are numerous methods of making baklava, layers of filo in a baking dish and cut into slices, rolled up in single sheets of filo or several sheets and cut up in slices. The method is up to you. Rolling them up singly is little more time consuming but they make nice neat little parcels, which can then be sliced into smaller portions. Chop the macadamia nuts by hand if you can, otherwise if they are ground in a food processor, the nuts tend to clump together and the mixture becomes too moist.

20 sheets of filo pastry

1 cup macadamias – finely chopped

125g unsalted butter - melted

1 tsp. ground cinnamon

¼ tsp. ground cloves

¼ cup sugar

2 tbs. rosewater

2 tbs. chopped macadamias

Syrup

½ cup castor sugar

¼ cup water

2 cloves

1 cinnamon stick

¼ cup honey

2 tbs. rosewater

1 tbs. lemon juice

Place ingredients for the syrup together in a small heavy-based saucepan. Stir until sugar has dissolved. Heat up slowly and simmer for approximately 15 minutes, or until syrupy. Remove the cinnamon stick and cloves, discard. Set syrup aside and allow to cool. In a bowl, place the chopped macadamia nuts, ground cinnamon and cloves, sugar and rosewater. Combine well.

Preheat oven to 180°C. Lay out the sheets of filo pastry and cover with a damp cloth to stop the pastry from drying out. Take one sheet of pastry, lay out horizontally and brush liberally with melted butter. Fold in half, from left to right. Brush with melted butter again. Place one tablespoon of the macadamia mixture onto the bottom third of the pastry, spreading the mix out to a 10 x 10cm remaining approximately 10cm from each edge (visually divide each pastry into thirds, horizontally as well - the mix goes on the centre third). Bring in each side and fold over the area with the mix on it. Brush each side with butter again. Then roll into a stubby cigar shaped parcel. Do not roll up too tightly otherwise the rolls may burst while baking. Place each roll onto a lightly buttered baking dish, leaving an inch space between each roll with the seam of the roll facing down. Repeat with each sheet of pastry until they are all rolled up.

Place in the oven and bake for 30-45 minutes, until golden brown. Allow to cool for 5 minutes and then brush with syrup and sprinkle with the extra nuts.

Serve sliced into thirds.

to Woombye's fruit growing list. Although not generally thought of as well suited to the tropics, they have been grown successfully by a number of growers. Figs are a delicate fruit; easily torn open, they reveal a fibrous maze of fleshy edible seeds that has the most exquisite, gentle aroma. Figs are sensational as part of an antipasto selection, marrying well with cheeses, smoked meats, alongside a glass of good red wine. Figs ripen from January to May and one Woombye farm in particular will generally display a sign out the front of the property when they are available, so keep a look out for that.

Figs

Figs are a succulent and sweet fruit that are high in fibre and calcium. A beautiful companion to cheese and wine, they can be bought from the farm gate just down from the Big Pineapple when in season.

Look out for:

- The sign that says 'Figs for Sale!'
- Fig jam

An initiation to the Sunshine Coast would not be complete without having been introduced to the most evocative of fruits, the mango. Mangos have a taste like no other fruit, delicately sweet and juicy, the vivid orange flesh almost melting in your mouth, it is hardly any wonder that this tropical favourite is revered by so many. Mangos are native to India and Malaysia and eventually made their way to our shores to be cultivated throughout the tropics. The splendorous sprawling trees grace the yards of many established homes and once relieved of their abundant cargos are made into luscious jams and chutneys to be enjoyed the whole year round. **Sunshine Tropical Fruits**, based just out of Nambour, became the first commercial processor of mangos in Australia and continues to supply a significant portion of mango product to the food service sector to be processed in a variety of ways. Along the way, Sunshine Tropical Fruits has collated quite a repertoire of processed food products to include almost every berry and locally grown fruit available. They now boast over 70 varieties of jams under the Mother's Kitchen and **William's**

brand names. **The Mother's Kitchen** range of jams and preserves has the most wonderful assortment of flavours, with a special focus on tropical flavours such as Mango and Macadamia or Passionfruit and Macadamia. Relishes and chutneys are likewise; the Mango and Ginger Chutney and the Pineapple Relish are just two of the many unique flavours on offer.

The Strawberry is the next unforgettable fruit grown widely on the Sunshine Coast. Strawberries are surely the most ambrosial of fruits. They are a seductively, fragrant red berry that has developed from early history's tiny wild woodland fruits into the plump, luscious berries we see today. Strawberries are grown commercially in various regions on the Sunshine Coast each region developing varieties that are most suited to it. Almost half of Australia's strawberry production comes from Queensland, the warmer winter and spring climate providing perfect growing conditions, continuing strawberry supply throughout the colder months to the Southern states. The freshest ones are the ones you pick yourself, a past time most local growers encourage. **Strawberry Fields** is one of those growers. Located at Palmview, not far from the Ettamogah Pub. The farm is open from July to November, where the strawberries can be picked from the field, or bought in ready, made up punnets. There are many local growers who allow the more enthusiastic among us to pick their own, so hunt around and make a day of it. There is nothing more rewarding than making your own strawberry jam, despite the fact that there are wads of producers that have gone to the trouble for you. Some of them even add a twist of ginger or chilli.

Strawberries

The popularity of strawberries is legendary, symbolic of love and romance, they are the most ultimate of fruits. In season throughout summer, most growers will allow you to pick your own berries straight from the fields.

Look out for:

- Choc-dipped strawberries
- Strawberry jam
- Strawberry chilli sauce
- Strawberry sorbet
- Strawberry yoghurt

Strawberries & Chocolate Fondue

500g punnet strawberries

4 tbs. castor sugar

1 tbs. Drambui

1 cup cream

1 nougat bar

500g dark chocolate

250g marshmallows

1 tbs. Drambui - extra

Wash, hull and halve strawberries. Place in a bowl with sugar and Drambui. Toss lightly and set aside.

Place all remaining ingredients into a small heavy-based saucepan and heat gently until mixture has melted, stirring constantly.

When ready to serve, place in a heatproof dish and place over a fondue burner. Dip strawberries into the chocolate sauce and eat!

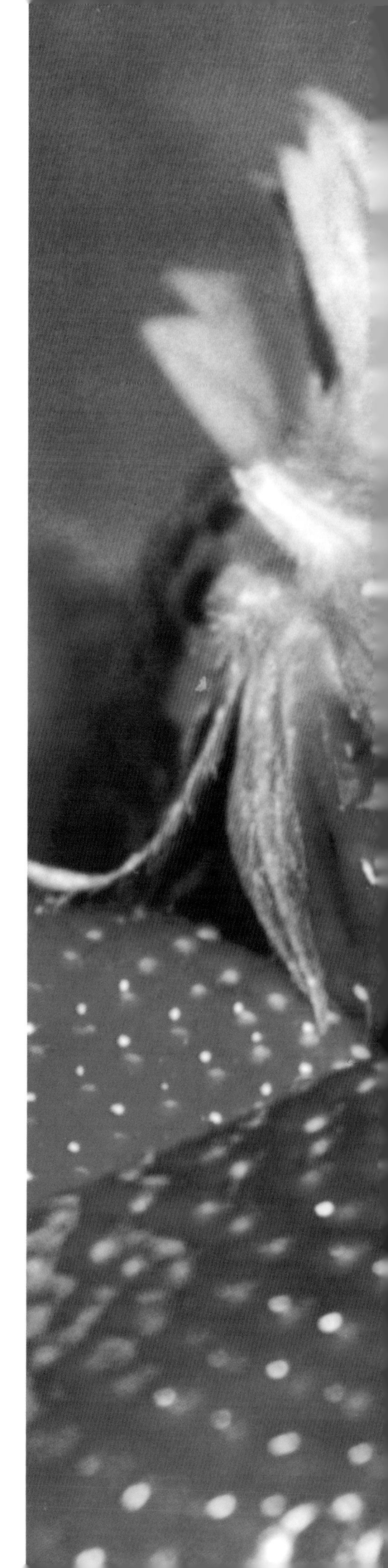

BLACKALL RANGE

Blackall Range

The Blackall Range is the panoramic prize of the Sunshine Coast's magical hinterland. The spectacular mountain range links the towns of Maleny, Montville and Mapleton, the scenic winding road travelling along the spine of the range, sea views to the east and rolling green pasture and national parkland to the west. Approaching Maleny from Landsborough, the traveller is treated to glimpses of the evocative Glasshouse Mountains, each new curve in the road providing yet another breathtaking moment of nature at its best. Mary Cairncross Park holds the most picturesque vantage point of all – a place you have to remember to come back to once you have stocked up on local goodies, to picnic at later on in the day.

In Maleny, self-sufficiency and community mindedness is at the heart of this town in fact the whole region. Bakers, butchers, even the local supermarket, support local industry first and foremost. **The Maple Street Cooperative** is just another example of the community's insistence on nurturing a healthy, wholesome and self-sufficient community.

Maleny as a town has a wonderful feel. It has an eclectic mix of café and gallery style gift shops and locals are friendly and welcoming. The most recent addition to Maleny's gourmet world is **Maleny Cheese**, a boutique dairy goods manufacturer just out of the town itself. The Cheese Stop allows visitors to watch the cheese making process in action as well as the opportunity to try some of the wonderful cheeses they make. Their Camembert and Brie are especially good.

There are a few wonderful gourmet circuits branching from Maleny. You can travel through the Obi Obi to Kenilworth, or through the Blackall to Mapleton, but my favourite and one of the most scenic routes is through Bellthorpe to Woodford. The first time I travelled to Bellthorpe was an absolute thrill. The road snakes through dense bushland up an incredibly steep incline to one little open pocket where the view of the Glasshouse Mountains is just magic. Just about everyone that ventures past here for the first time, stops to take a photograph. The Bellthorpe region is lush and rich and slightly cooler conditions have made it perfect for the breeding of deer. Although none of the breeders are open to

MALENY CHEESE
Colin James
Cheeses
Deli-lines
Specialty Items
Fine Foods
Purveyors of Gourmet Products
Maleny Qld. Est. 2001

the public, though a few of them have made noises of the kind. In the future there may be some evolvement in farm style visits, but for now you will just have to settle for the natural beauty of the area and a view of these spectacular animals from a distance. Roadside stalls are profuse with locally grown fruit and vegies so a trip can be quite rewarding regardless. There are a number of butchers on the coast that do deal with venison, so it is worthwhile making a few enquires.

Another meat that has made quite a presence in the marketplace is that of Emu meat. A few local restaurants have been using this flavoursome meat, and Sydney and Melbourne have not been afraid to experiment with some really interesting results.

Emu farming is still a relatively young industry, not without financial teething problems, although those that have stayed in the industry have made a conscious effort to focus on emu products as a whole rather than just the meat. The oil and leather are considered the most valuable. Generally a very tender meat, emu meat it is at its most flavoursome served rare, smoking also being a very popular cooking method. From a nutritional perspective, the meat is a finer textured meat than beef, low in fat and cholesterol and much higher in iron.

Turinga Emu Products, one of the more active breeders in the region is based just out of Maleny and have invested a great deal of time in educating the public. They have stalls at a lot of the farmers markets where emu sausages and a variety of other products are available for sampling in a hope to demystify the meat. Being more than a little sceptical at first, I was pleasantly surprised; the flavour

Venison

Venison is a flavoursome game meat. Its prime cuts need minimal cooking to exploit its fabulous flavour, though the lesser cuts come into their own slow cooked to make hearty pies and braised dishes. Butchers do stock venison by special request

Look out for:

- Venison fillets
- Venison pies
- Venison carpaccio

is extraordinary. There are some butcher shops that stock emu meat; again, like venison it is a matter of a few enquiries. Available for purchase are fillets, sausages, smoked meats and salamis as well as a range of emu pies.

Emus

Emu meat is a healthy red meat that is high in iron and low in fat and cholesterol. The fillets are at their most tender when served rare to medium. There are a number of butchers on the coast that stock emu meat by special request.

Look out for:

- Emu fillets
- Smoked emu
- Emu sausages
- Emu salami
- Emu pies

At the Turinga property the breeding process begins with the incubation of the eggs. As a matter of interest, in the wild the male actually sits on the eggs, while the females release a ritual, rather haunting, guttural 'drumming' sound that can be heard from quite a distance. Cute stripy chicks emerge from their shells after approximately 55 days and are then nurtured for 2 years before they are ready for the market. At the moment Turinga's focus is largely on the use of the luxurious emu oil in their own developed skin care products, which is proving a very viable and sought after additional by-product. Turinga is exploring the option of farm visits, which hopefully happens. Emu farming is a fascinating industry that will no doubt, have a huge future once all the myths are quelled.

While on the subject of Australian native produce, the native fingerlime is another emerging native industry in the Maleny region. Fingerlimes are a citrus like fruit, which is, as yet, relatively undiscovered except by some of the more creative chefs that have explored some of the intense flavours indigenous foods have to offer.

Fingerlimes are finger shaped fruits that dangle like Christmas decorations from small, prickly evergreen trees. The fruits themselves have brilliant magenta leathery skins, which when sliced spill out dozens of tiny tinted 'rainforest pearls' that are bursting with tangy intensity not unlike that of the conventional lime. **Coolbooroo**

Pippi's & Fingerlimes in Rosé

On the Noosa Northshore, pippis are a dime a dozen, all you need to do is dig them out of the sand and sit them in a bucket of fresh water for 24 hour to give them time to spit out the grit and sand. However should the inclination to forage the Northshore to Rainbow Beach fail you, then mussels or clams, which are available at some of the local seafood shops, as an alternative to pippis, is the way to go.

1kg pippis

½ bunch spring onions - sliced

1 tbs. fresh ginger - grated

1 clove garlic - crushed

1 chilli - finely chopped

2 fingerlimes - pulp (4 tbs.)

1 bunch coriander - chopped

salt and pepper

1 cup Rosé

2 tbs. olive oil

In a large stockpot, heat olive oil. When hot, add spring onions, ginger, chilli and garlic and sauté briefly. Add Rosé and turn up the heat. When boiling vigorously add the pippis and put on the lid. Shake the pot about at regular intervals so that the pipis all get even heat. As the pippis start to open remove them with tongs onto a warm platter. Discard any that don't open.

When all the pippis are out of the liquid, reduce the liquid for 1 minute and add the finger lime pulp, stir in and season with salt and pepper. Pour liquid over the pippis and sprinkle with coriander.

Serve immediately. Serve 4-6

Bush Foods look set to explore a multifaceted future in the bush food industry with plans for preserves and other value add products utilising the fingerlime. I love fingerlimes. The pearls introduce a tartness and tanginess in the way of flavour to dishes that they are incorporated in. Possibly a fingerlime addict, I initially almost drove my family mad in my over-enthusiasm to incorporate the pulp into nearly every single dish we ate…. well, at least I now know what works and what doesn't! What did work was the compatibility of fingerlime with fish and seafood and they almost beg to be stirred through steaming rice and white wine sauces. The possibilities are endless and what is even better is that the pulp keeps extremely well frozen in ice cube trays to be used throughout the year. Experiment with this fruit when available, as the results are sensational, sweet or savoury.

Fingerlimes

There are only a handful of fingerlime growers, so they are not easy to find. They make an appearance in autumn at some of the markets and the more boutique style fruit and vegetable stores – try the Maple St Co-op.

Look out for:

- Fingerlime marmalade
- Fingerlime nectar
- Fingerlimes on restaurant menus

Have you often entertained the odd wave of whimsy for a career as a 'moonshiner'? Wild thoughts of distilling one's own spirits with a rickety old homemade still have been left unfulfilled but the concept continues to intrigue. A few years back I came across a 'Wild Crafter of Rainforest Liqueurs' and realized that it is not such a whimsical career choice. If you take away the rickety old still and the 'moonshining' label.

Rainforest Liqueurs is based just out of Maleny – yet another confirmation of the regions commitment to native industries - they use the aromatic fruits and leaves of native plants to make exquisite liqueurs. Sandpaper figs and blue quandongs are used to make the

Ke-ril and Cooloon Liqueur, exotic fruity liqueurs, while the leaves of the native aniseed or cinnamon myrtle readily impart their subtle spicy tones in the Anisata and Myrtifolia Liqueur. Rainforest Liqueurs indulgence list includes over 15 flavours, many of them produced in limited numbers due to the scarcity of the raw products, taking special care not to deplete or affect the natural regeneration process through excessive harvesting. The liqueurs are an unusual, yet sensational take on the bush food concept and really worth looking out for.

Bread is a culture. In fact the whole bread making process could be defined as one. From the actual culture that bread is leavened from, through to the intense community fabric that keeps the bread making process alive. The mass production of bread has seen bread loose its identity as far as the baking ritual goes, but independent bakers are slowly rediscovering these rituals and fortunately, the culture is far from gone. There are many bakers on the Sunshine Coast, too many to pay homage to them individually. Many restaurants bake their own signature breads, while there also seems to be a resurgence in traditional wood-firing techniques as opposed to the less romantic option of the industrial oven. Markets are a great place to buy some of the more unusual breads although a lot of the local bakeries should not be underestimated. Nearly every town, no matter how small, has a bakery or baker that bakes its own delicious specialty bread.

Pineapples

Pineapples are grown in abundance throughout the Sunshine Coast. They are rich in vitamin C and have an unmistakable sweet fragrance. The Big Pineapple and Yeltukka Pineapple Plantation cater for the more curious, though pineapples are available almost everywhere when in season.

Look out for:

- Choc dipped dried pineapple slices
- Pineapple jam & relishes
- Pina colada conserve

A bakery tends to become the economic heart needed for a sustainable community and the place where this is probably at its most apparent is at Crystal Waters Permaculture Village just out of Conondale. Crystal Waters is an eco-friendly village developed on the principals of permaculture where residents are encouraged to work together and in harmony with nature.

The **Crystal Waters Bakery** would be one of the village's success stories so far and hopefully not the last. It is a fabulous experience watching the baker at work in this bakery. Mounds of dough are laid out on a solid timber bench and then kneaded and manipulated into the various loaves ready for baking and all while the baker catches up on community activities as locals go about their work. This is how bakeries used to be before plastic wrapping and supermarket convenience enraptured us all.

Crystal Waters Bakery focuses on sourdough breads. They are made with organic flour and a natural leaven instead of baker's yeast,

Bread

There is nothing quite like crusty bread fresh from the oven. Sour dough and spelt breads are available from the more boutique style bakeries although all breads are fabulous fresh.

Look out for:

- Spelt breads
- Wood fired breads and pizzas

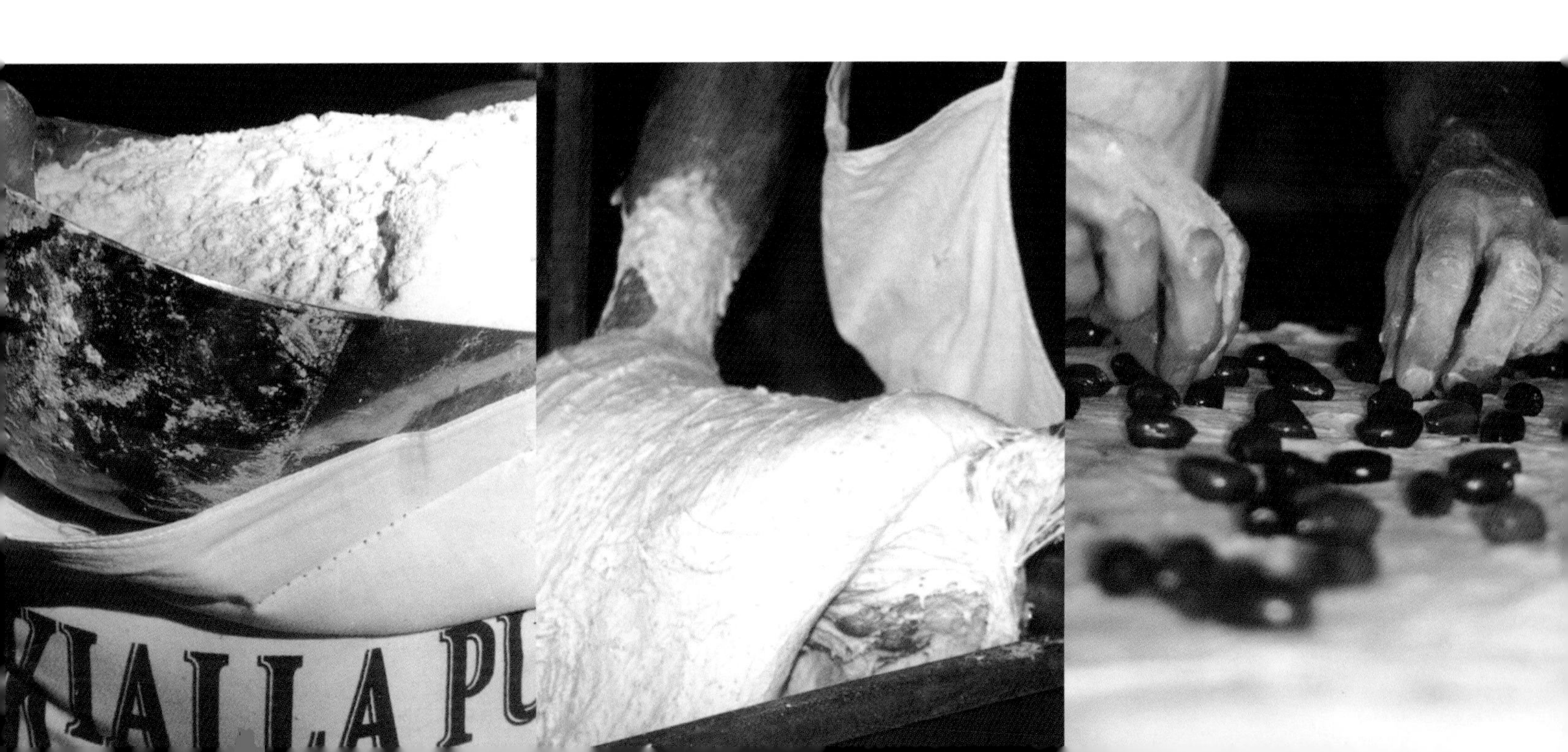

producing a culture that results in delicious bread that is easily digested. The wood firing usually takes place at night and has become something of a community ritual, visitors keeping the bakers company, coming and going through the arduous hours of baking well into the night to provide fresh bread for the next morning's markets.

The village welcome visitors to its information centre where tours can be organised if you would like to know a little more about permaculture or the village itself – courses and weekend workshops also run throughout the year. The may have a bread-making workshop, so ask.

The Sunshine Coast's most spectacular vantage point would have to be from the incredible picturesque stretch of road that links Maleny and Mapleton. The mountain roads roll out into a pristine landscape

Smoked Goods

Continental smoked meats and sausages are a popular addition to antipasto platters and picnic baskets. Specialty delicatessens do stock some local smallgoods and there is Franz's Smallgoods in Caloundra that allow you to buy direct.

of farmland and natural bush to a view of the coastline's stunning beauty.

Confectionery

There is no shortage of confectioners on the Sunshine Coast. Between the fudge, candy makers and chocolate crafters the sweet tooth is amply catered for.

Look out for:

- 'Bliss' nougat
- Handmade chocolate truffles
- Candy
- Fudge
- Meringues
- Macadamia bars

Magical Montville, as it is known, is situated at the top of this glorious range. It has a distinctly European 'alpine' ambience, timeless tradition further enhancing the magic that already exists. Montville seems to be the burning flame of the artisan world, galleries and craft shops igniting either side of the main road, with handcrafted and hand-fashioned ware unlikely to be seen anywhere else on the coast.

Sugary indulgences also abound and have managed to prise themselves a position into the psyche of Montville itself, sweet-tooths clamouring to indulge in at least one of the towns hypnotizing display of sugary treats. **Chocol'art** touches the heart of the most faint-hearted chocoholic, visually and aromatically, as does **Aunty Maureen's Fudge Factory**. In an instant, vanilla and chocolate dance exotic dances with peppermint and butterscotch, mingling a melody of flavour and fragrance. Hand-cut blocks of fudge are neatly stacked and on display, colours just screaming out, to even the most passive of shoppers.

Montville's winding paths lead from one treasure to another, **The Sunshine Candy Kitchen** is next. Humbugs, acid drops, rosy-apple candy – the names are almost as colourful as the candy itself. Rainbows of sugary twirls and swirls are an endless kaleidoscope of vivid patterns and shapes, the magical result of a couple of artful candy makers and Montville's old world charm couldn't complement the candy making concept more. The traditional store recaptures childhood memories with its old fashioned candy with flavours that

almost explode in your mouth. Lollypops, fruit rocks, bulls eyes, koala droppings….. the list goes on.

Bananas

Bananas are one of Australia's most popular fruits. They are virtually fat free and considered a high-energy food. There are a number of growers experimenting with some of the more unusual varieties such as plantains, which we may see more of the future.

Look out for:

- Red Daccas
- Plantains
- Lady fingers
- Banana butters

The candy making process is as spectacular as it is fascinating. Vibrant combinations of colour and flavour are twisted and stretched on a hook, the candy taking the appearance of spun silk as it is plied into coils of various hues. Stunning images are created in the centres of these flavoured rocks and you are left wondering how on earth they got there. There is no doubt about the complexity of the designs and the intricate planning that took place to create them.

Demonstrations are held in store most days, where lolly lovers are glued, noses almost touching the glass in complete awe as the lollies are created. It is such a lot of fun for both adults and children, the adults reliving childhood memories while the kids don't really care just as long as they get to walk out with some of their newly found favourites. If you don't get the opportunity to go to Montville there is another candy maker, **Cane and Able** in Coolum. Cane and Able have a range of really funky candy as well as the more traditional favourites.

Before leaving Montville, one final experience is that for the coffee lover. Coffee lover or not, inhaling the aroma of a bag of freshly roasted coffee is a heavenly experience. Western cultures have been enjoying this tradition since coffee swept across Arabia from Ethiopia, where it originated centuries ago, into Europe and the homes of coffee loving Westerners. It has resulted in coffee becoming the cornerstone to our society, kick-starting days, offering time out and strengthening the socials bonds for millions of people all over the world – all this in a cup!

The plant on which the coffee is grown, is itself lush and green. It is fruitfully lined with burgeoning red berries when at its peak. When harvested they are fermented and then roasted to unleash an intensity of flavour that can only truly be captured in a fresh pot of ground and filtered coffee – instant coffee the greatest sacrilege for the coffee connoisseur.

The home to **Montville Coffee** couldn't be more tranquil. Over 4000 Arabica coffee trees have flourished amongst native trees and shrubs, nature's shield from the elements. Seek out the pleasure of attending a lunch at the family tree-house, a rustic little building perched on the edge of the property overlooking the Baroon Pocket Dam, complete with sugar glider and fresh rainforest air. Freshly brewed coffee never tasted as good. That, is part of the coffee experience. It is time out, a savouring moment, it enhances olfactory senses and suddenly you are breathing in the aroma, inhaling the scents all around you, the fresh air, the sea breeze… Montville Coffee produce various blends, supplementing their stocks with organic beans from New Guinea and East Timor, but are hoping that with time their own crops will provide all they need.

Coffee

Freshly roasted coffee doesn't get any fresher than locally grown, some growers offering the additional bonus of their coffee being organic.

Look out for:

- Montville Coffee Organic blends
- Decaffeinated
- Espresso Coffee Liqueur
- Mocha Ice cream
- Choc-coated coffee beans

Ok, so now you have filled up on sugary treats, had a coffee, stopped for cheeses and fresh crusty bread and now you need a good wine to complete the picnic basket. The wine industry on the Sunshine Coast is still a relatively young industry, however there have been a remarkable number of top quality wines being produced, in what will no doubt be a prosperous industry in the future. There are a number of wineries scattered throughout the Maleny to Mapleton

Bananas & Black Sticky Rice

This is a great dinner party treat. They can be assembled in advance and then steamed prior to serving in stacked bamboo steamers.

3 cups water

½ cup black sticky rice

¼ cup palm sugar

1 cup coconut milk

2 bananas

½ cup water - extra

1 vanilla bean

¼ cup coffee beans

12 banana leaf squares

- 20 x 20 cm

12 lemongrass leaves or

toothpicks

Soak rice in water overnight, drain and rinse. Put the 3 cups of water in a saucepan with the rice and bring to the boil. Boil for 30 minutes or until tender. Drain and rinse again. Set aside. Mix extra water with sugar in another saucepan until dissolved. Add vanilla bean, which has been cut in half lengthways and simmer. Add coconut milk and then stir in rice. Simmer on low heat until the moisture has been absorbed by the rice. Add the coffee beans. Remove from heat.

To prepare banana leaves either soak momentarily in boiling water to allow leaves to be more pliable. Lay them out. Cut bananas into thin slices and divide the slices between the leaves, laying them in the centre of each leaf. Place 2-3 tablespoons of rice mixture on each leaf and then fold the leaves bringing each side in to the centre. Bind parcel with the lemongrass leaves or secure with a toothpick.

To steam, use a large bamboo steamer that has several layers and place a few parcels in each layer and steam for 5 minutes. Serve while hot.

region, beginning with the **Maleny Winery**, before you enter the road through the range. Panoramic views are very much a part of the wine experience and most of the cellar doors have capitalised on this fact. Each winery sits in amongst a picture pretty setting of rambling grapevines and sprawling scenery. The industry itself is working together to promote wine and food trails to encourage locals and visitors to visit the individual wineries either independently or via an organised tour which currently ferries revelling wine enthusiasts to and from their chosen destinations. Once at the wineries, wine experts happily discuss the various wine blends along with the wine making process itself. They offer for tasting and purchase their own signature blends. Light meals are also available, focusing on local produce, especially local cheeses, wine's exceptional partner in crime. Montville has it's own **Settlers Rise Winery**, stunning sea views visible from the decks that over look the vineyard. A little further along the range road is the **Barambah Ridge Faxton Cellar and Vineyard** an equally picturesque winery. There is also the **Little Morgue Winery** just out of Yandina and the **Noosa Valley Winery** half way between Noosa and Eumundi, fairly new to the industry, but offering top quality wines none the less. Then of course there are the wineries previously mentioned, **Moonshine Valley Winery**, **Eumundi Winery** and the **Kenilworth Bluff Winery** – enough of a selection to celebrate any harvest in true bacchanalian style, although wine connoisseurs are a little more refined these days….

Wine

Local wineries are producing some sensational wines. The beginning of each harvest is usually signified by a string of wine festivals and harvest celebrations providing the perfect platform for each winery to introduce its latest vintage.

Look out for:

- Chardonnays
- Semillons
- Shiraz
- Verdelhos
- Rosés
- Grappa

Pumicestone Passage

Adventure is in the discovery and one continues to find hidden pocket and alcoves, each one yielding yet another secret bounty of some sort. Toorbul was another one of these hidden pockets and one of my final destinations. It is a magical little coastal town, nestled into the banks of the Pumicestone Passage, the body of water that divides Bribie Island from the mainland. It is reminiscent of the way coastal towns used to be, radiating a peaceful ambience as shimmers of water gently lap up to the shore, while a family of resident kangaroos graze calm and disinterested along the roadside, the tide, as it recedes, exposing a treasure of its own.

Toorbul Point has long been the home to the oyster, a curious mollusc with a rich history of gluttony and intrigue. There is no in betweens with oysters, either you like them, or you don't, and if you do, then the lust for them courses in your veins. As a passionate oyster connoisseur, the hardest choice you will ever have to make whilst dining out is whether to be satisfied with a paltry ½ dozen or give in to the urge to devour the full dozen!

Queensland led the oyster industry in Australia in the early 1900's, harvesting from plentiful natural oyster beds throughout the Pumicestone Passage. Sadly, this almost led to a depletion of the natural beds. Legislation was introduced in order to preserve and regenerate these beds. This saw farmers securing leases where they employed a stick and tray method, which was not without it's own set of problems. However the industry continues to thrive, be it on a much smaller scale.

It is interesting to note that the aboriginals that had lived in this area may have shared this passion for oysters, by the fact that one of the creeks is named Ningi Ningi Creek – Ningi Ningi was the name given to the oyster by the indigenous community.

Toorbul Point Oysters have farmed the waters of the Pumicestone Passage since the early 1960's although over the years their leases have been reduced to a much more manageable size.

The season begins in August where clumps of small Sydney Rock Oysters are bought in. They are placed on racks where they then grow to a plump edible size. They are harvested up until late December, removing the last oyster just in time to beat the full summer heat and the risk of spoilage and disease.

The oysters are sorted and then taken by punt to the processing facility on Bribie Island, where they are shucked and sorted further, according to size. Some are bottled, while others remain in their shells to be presented on dinner plates all over the Sunshine Coast – fresh from waterways that are natural and untainted.

Bribie Island is also the home to the Bribie Island Aquaculture Research Centre. The centre was established by the Department of Primary Industry to encourage the development of local aquaculture industries as well as playing a significant role in technological development and research. Current research programs include the potential for the farming of eels, development of live transport for the Kuruma Prawn as well as research into the cultivation of mud crabs, Moreton Bay bugs, scallop regeneration and native fish species.

Oysters

Freshly shucked oysters have an exquisite salty sea taste that makes it hard to stop at just one. There is, a small oyster plot in the Maroochy River although most of what we consider local oysters, come from further south, from the Pumicestone Passage.

Look out for:

- Fresh local oysters!!!
- Oyster Shooters
- Oysters Kilpatrick
- Oyster Bars

OYSTERS WITH FINGERLIMES & SALMON ROE

Oysters with Fingerlimes & Salmon Roe

1 doz. oysters

1 tbs. salmon roe

1 tbs. fingerlime pulp

½ tsp. ginger - freshly grated

1 clove garlic - crushed

1 tbs. olive oil

salt and pepper

In a small dish, combine roe, fingerlime pulp, ginger, garlic and olive oil. Season with a little salt and pepper and set aside for ½ an hour to allow the flavours to develop. When ready to serve, spoon dollops of the mix onto the oysters and eat immediately!

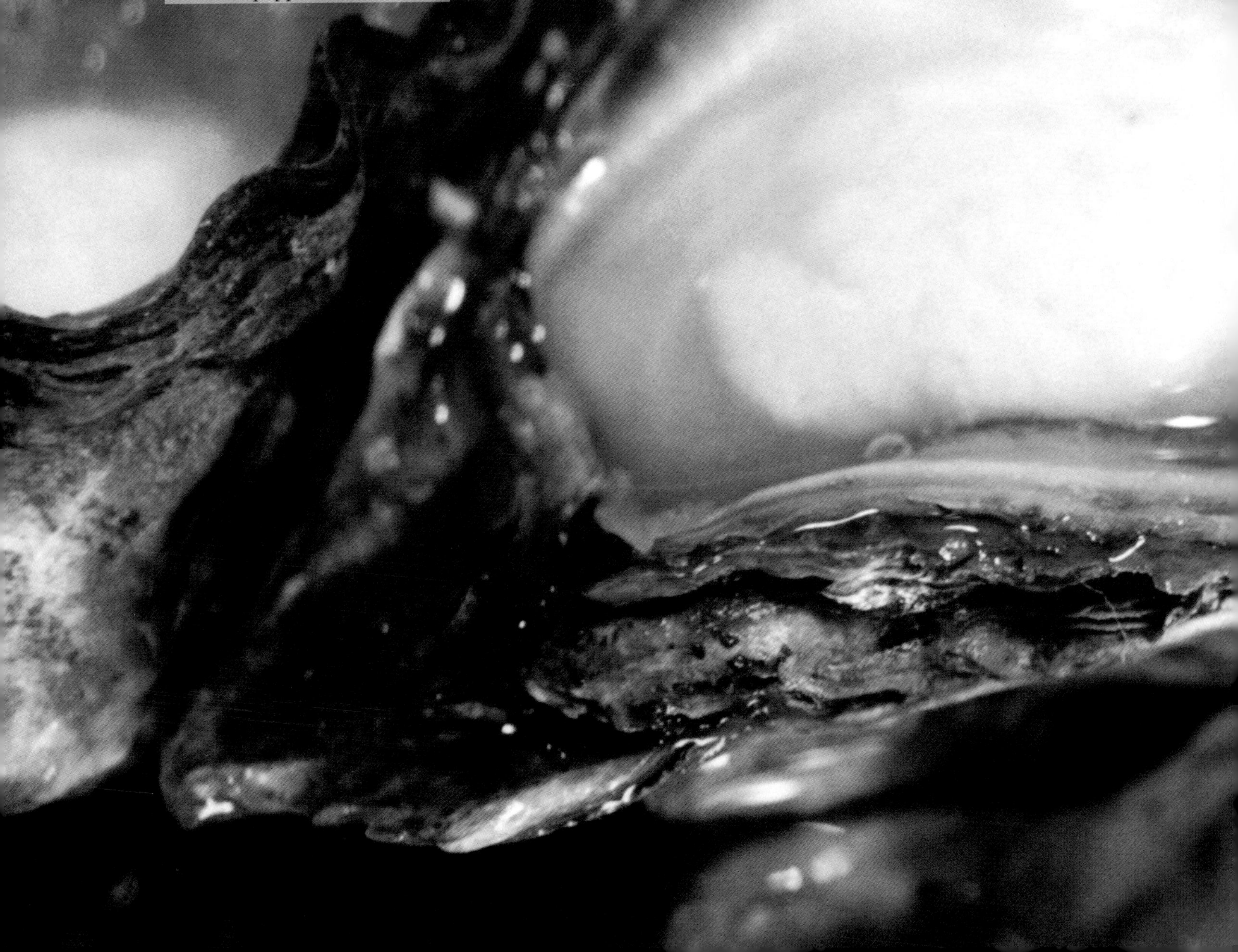

Carpetbag Steak

Carpetbags Steaks are a wonderful combination of flavours, the oysters, a plump and juicy surprise in the centre of the most succulent cut of beef. The hardest thing to explain is the cooking time of a steak. It is something one feels, rather than does, and everybody has their own preference as to whether it should be rare or well done. The thickness of the steak makes a difference too. A few minutes on each side and then left to rest for a few minutes more. I ask my butcher to specially cut my steaks nice and thick – 2 inches, but the thickness is up to you. Don't fuss around with complicated sauces. This sauce is so basic and is just to provide an extra bit of moisture and flavour, so feel free to substitute the sauce with a favourite white sauce recipe of your own.

4 thick slices of Rib Eye Fillet Steaks

3 dozen oysters

8 slices Prosciutto

1 bunch broccolini

1 bunch bok choy

1 cup fresh peas or snowpeas

Prepare steaks by slicing a pocket into the eye fillets, being careful not to cut all the way through. Line the cavity with one slice of prosciutto. Place 4 oysters between the prosciutto. Secure the pocket with 2 or 3 toothpicks so that the oysters don't slide out during cooking. Set aside.

Prepare the vegetables. Slice the spring onions into 10cm long pieces. Trim the ends off the broccolini and halve. Cut the stalks off the bok choy and keep them separate to the leaves. Divide the parsley into little sprig segments. Trim the snowpeas or prepare the peas. It is important to have everything prepared as once the cooking process begins everything will proceed very quickly. Heat the oven to 180°C and then turn off.

4 spring onions

1 clove garlic – sliced finely

few sprigs of parsley

1 tbs. butter

1 cup cream

¼ tsp. wasabi paste

1 clove garlic – crushed

1 tbs. olive oil

salt and pepper to taste

toothpicks

Heat a saucepan full of water and bring to the boil. Heat a little olive oil in a heavy-based frying pan. When hot, place the steaks in the frying pan and cook on each side for the desired time – remember oysters are nicer juicy!! In a separate pan, fry the remaining prosciutto until crispy and then drain on absorbent paper and put on a dish in the warm oven. Do not turn the oven on, we just want to keep the strips warm until it is ready to serve. When the prosciutto has been removed from the pan, remove excess fat and then pour cream into the pan. Do not clean the pan as the prosciutto residue imparts its flavour into the cream. Turn up the heat, bring cream to the boil, add crushed garlic and then allow to continue simmering until the cream starts to thicken. Add wasabi paste and salt and pepper to taste and set aside. When you turn the steak over, start cooking the vegetables. Put the broccolini and the peas/snowpeas into the simmering water and blanche for 30 seconds, then drain.

When steaks are cooked to perfection, place on a dish and sit to rest in the oven. It is worthwhile to put the serving plates in there as well so that they are nice and warm when the meal is served up. Melt the butter in another saucepan and sauté garlic slices, bok choy stalks and spring onions for one minute. Then add the bok choy leaves, blanched broccolini, and snowpeas/peas and the parsley. Toss and coat with butter. Just before serving add the remaining oysters to the cream sauce and heat long enough to heat through.

When ready to serve, place a mound of the greens on a plate and place the steak on top. Spoon over the cream sauce, making sure each portion has a few oysters in the sauce. Prop a prosciutto strip on top of each steak.

Serve immediately with a wedge of lemon. Serve 4.

...on a final note

...and so the journey ends, but not really, because the Sunshine Coast is forever evolving and new industries are constantly being created. You can go to the same market three weeks in a row and still discover something new.

When I started on this book, it was all about encouraging people to support their local industries, it doesn't really matter which ones in particular, just if there is a local alternative, try it and if it is as good or better, then support it! Sadly in the process of finishing this book, there were a number of spectacular producers that were forced to shut down their operations simply through the lack of support from the community or the difficulty in being producer, marketer and retailer all in one and every time this happens I feel a great deal of regret.

Now, I have tried to be impartial while collating the information for this book and while I have my favourites, it is not really about what I like. For me, as I mentioned earlier, it is all about the discovery, going somewhere new, finding something new, coming home and experimenting with it and then sharing my discoveries with others who are equally interested in these fabulous gourmet industries. There are many producers who I haven't included simply because there is not enough space in any one book to mention them all and I would like to stress to readers that their omission is due to space restrictions only. I would also like to thank all the producers mentioned in this book who willingly gave their time and knowledge and an extra special thanks to those who sent me home brimming with goodies to play with.

On that final note, I would like to encourage everyone to support these local industries. They are our community's pioneers, they provide work and opportunities for our younger generations and provide an economic backbone that feeds the community when times are times are tough and the reward, well, there is no greater reward than eating produce that is fresh, natural and above all, locally produced.

Producers

(Open to the Public)

Aunty Maureen's Fudge Shop
Main Street
Montville
Ph. 5478 5455

Barambah Ridge Flaxton Cellar & Vineyard
313 Flaxton Drive
Flaxton
Ph. 5478 6666

Belli Bamboo Parkland
1171 Kenilworth Road
Belli (via Eumundi)
Ph. 5447 0299

Big Pineapple
Nambour Connection Road
Woombye
Ph. 5442 1333

Coolabine Farmstead Goats Cheese
PO Box 185
Kenilworth
Ph. 5446 0616

Crystal Water Permaculture Village
Kilcoy Lane
Crystal Waters
Ph. 5494 4620

Eumundi Markets
Memorial Drive
Eumundi
Ph. 5442 7106

Eumundi Noosa Milk
331 Kenilworth Road
Eumundi
Ph. 5442 8890

Eumundi Vineyard & Winery
310 Memorial Drive
Eumundi
Ph. 5442 7444

Franz's Smallgoods
15 Industrial Ave
Caloundra
Ph. 5491 3311

Garnisha Curries & Gardens
44 Hatch Road
Boreen Point
Ph. 5485 3386

Kenilworth Cheese Factory
45 Charles Street
Kenilworth
Ph. 5446 0144

Kenilworth Bluff Wines
Bluff Road
Kenilworth
Ph. 5472 3723

Maleny Cheese
Clifford Street
Maleny
Ph. 5494 2207

Maple Street Co-Op
37 Maple Street
Maleny
Ph. 5494 2088

Moonshine Valley Winery (7 Acres Winery)
374 Mons Road
Forest Glen
Ph. 5445 1198

Noosa Farmers Markets
155 Weyba Road
Noosaville
Ph. 0418 769 374

Nutworks
Pioneer Road
Yandina
Ph. 5472 7777

Settlers Rise Winery
249 Western Ave
Montville
Ph. 5478 5558

Strawberry Fields
Laxton Road
Palmview
Ph. 5494 5146

Sunshine Candy Kitchen
Shop 7, 171-183 Main Street
Montville
Ph. 5478 5022

Suncoast Gold Macadamias
Drummond Drive
Gympie
Ph. 5482 7599

Superbee Honey Factory
Tanawha Tourist Drive
Tanawha
Ph. 5445 3544

The Australian Nougat Company
4 Tallgum Ave
Eumundi
Ph. 5442 7617

The Ginger Factory
Pioneer Road
Yandina
Ph. 5446 7096

The Spirit House
4 Ninderry Road
Yandina
Ph. 5446 8994

Sunshine Coast Brewery
5/13 Endeavour Drive
Kunda Park
Ph. 5476 6666

Toorbul Point Oysters
9 Armitage Street
Bribie Island
Ph. 3408 2844

For more information on local producers see:
www.localharvest.com.au

INDEX

Photography Locations:

Page 5 – Noosa Granite Bay
Page 10 – Noosa River from the lookout
Page 11 – Noosa from the 'Coastal Track', Noosa National Park
Page 21 – Eumundi
Page 30-31 – Jondell Macadamia Plantation
Page 37 – Cooroy
Page 45 – Driveway to Belli Bamboo Parkland
Page 65 – Mt Coolum
Page 69 – Mooloolaba at sunrise
Page 72-73 – Mooloolaba waterfront
Page 93 – Glasshouse Mountains from Bellthorpe
Page 97 – Maleny
Page 113 – Bribie Island

Acknowledgements

There are always many little helpers who on a project such as this dedicate time and effort to bring it all to fruition. First of all I would like to thank all of the producers that generously gave their time and support especially to Rob and Ilka Clarke, Karen and Richard Barrett, Durnford Dart, Les Bartlett, John and Mary King, Cathy and Scott Shearer, Gary and Faye Kidman-Lewis, Tom Weidmann, David Groom, Andrew Kattenhorn, Michael and Julie Joyce, Andi Flower, Anthony Everingham, Shauna and Peter Wolfe, Tim and Claire Warren, Trevor Gough, Kathryn and Paul Lloyd, Martha Shepherd, David Haviland, Cheryl and Shane Grant and Peter Thompson as well as anyone who should be on this list that isn't.

There have also been a number of industry people who have given me invaluable advice and encouragement. Those are Julie Flanigan, Lyn Donaldson, Sue Koro and especially Danny Anderson and Selena Ross.

Then of course there are my wonderful family and friends. Thank you especially to my partner Don and son, Aaron for putting up with my constant distractions and food experimentation, for their praise for my successes and their silence on my failures!! Thanks to mum and dad for always, always encouraging me and for proving me with a wonderful childhood amongst the ducks and the geese and the sheep and the apples and the veggies. Thanks to my brother Richard, for all his help with my blasted computer. Thank you to all of my wonderful friends for their endless support and not forgetting who I am and especially to Anita who, without her help and pushiness this would never have been completed, not to mention the hours of endless scanning. Thank you sooooo much.

Finally and most importantly thank you to Charles Burfitt, my publisher for providing the final piece in the puzzle.